AIR OF DOUBT

Exposing the largest
and most costly scientific
error in history

FRANK A TINKER, PH.D.

www.airofdoubt.com

© 2020 Frank A Tinker, Ph.D.

For permissions contact:
ftinker@ airofdoubt.com

Eclipse Cover Photo From
Kitt Peak National Observatory
https://noirlab.edu/public/images/noao-4094/

Dr. Tinker's academic credentials include a *Summa Cum Laude* Bachelor of Science degree in engineering, a Master of Science degree in physics, and a Doctor of Philosophy in Physics. He has been awarded numerous patents in diverse fields including heat engine efficiency enhancements, analog parallel processing topologies for use in computing linear transforms, and components of a miniature actuator in worldwide use powering a life-saving artificial heart.

TABLE OF CONTENTS

Getting right to it, this is the global mean temperature for the last *420,000* years, or so. It also includes the next *100,000* years for good measure.

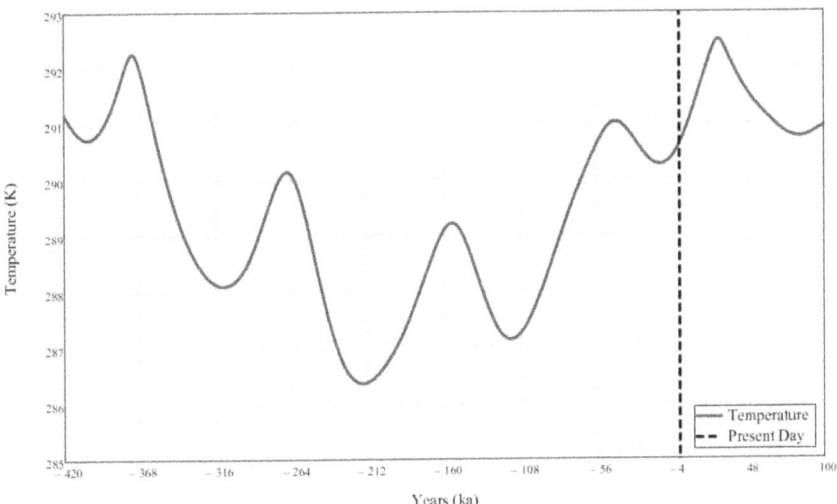

Climate scientists, however, are of the opinion that this is the global mean temperature for the last *420,000* years, or so.

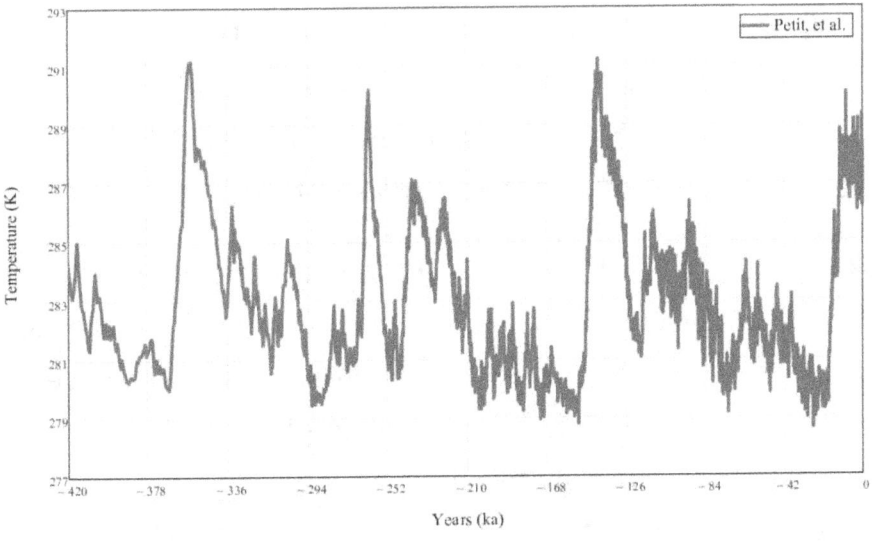

Wait, what?! How can that be?

Well, the truth is these two plots are very closely related and this is exactly how they are related.

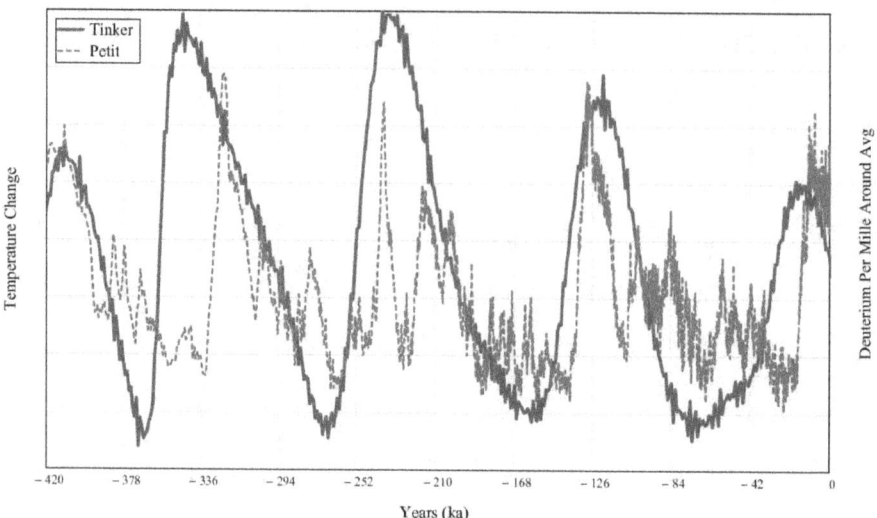

What this shows is the time rate of change of the global mean temperature (dT/dt) for the last *420,000* years. So, climate scientists have been assuming that data from ice cores and little creatures called "forams" are temperature proxies when, really, they are temperature-change proxies.

I have some good news and some bad news. The good news is that greenhouse gasses are not going to make the planet uninhabitable. The bad news is that we humans have spent about a trillion dollars over the last 50 years developing a science that, sadly, does not exist.

This book details where this all went wrong. In the process, it proves that Milutin Milankovitch (1879-1958) was exactly right in attributing global mean temperature changes to variations in Earth's orbit. It develops the correct mathematical relationship to describe that and, In the process, it reveals that Kepler's Third Law requires a minor adjustment to include the energy of an orbit. That adjustment resolves a long-known precession anomaly in Mercury's orbit that only General Relativity could before.

Buckle your seatbelts. It's going to be a bumpy...

Chapter 1 - Executive Summary

This chapter reveals that climate-science researchers have been mistakenly ignoring geothermal heat flow in their calculations of Earth's predicted surface temperature. The mistake was to combine solar flux and geothermal flux in a manner inconsistent with the superposition principle, a standard mathematical technique used when solving linear partial differential equations (in this case, the heat equation). Ignoring geothermal flux ($\sim 0.065 W/m^2$) results in a calculated surface temperature 33K less than it would be otherwise. This is confirmed both by measurement and by the Stefan-Boltzmann equation:

$$T_g = \left[\frac{0.065 \frac{W}{m^2}}{5.67 \cdot 10^{-8} \frac{W}{m^2 K^4}} \right]^{0.25} = 33K.$$

Properly combining geothermal and solar flux in a manner consistent with the superposition principle, results in the accurate determination of Earth's surface temperature using only those two heat sources. As such, no other contribution is necessary, or even possible. Since those two sources, alone, fully predict Earth's surface temperature, the Greenhouse Effect is superfluous. It is, in fact, shown to have been an invention necessary in order to compensate for the failure to include geothermal flux in the first place.

There is no argument in the climate-science community that geothermal flux is considered insignificant and, therefore, ignorable in the calculation. However, the analysis that draws this conclusion is inconsistent with the superposition principle. As such, ignoring geothermal flux is not just an admitted failure to respect the first law of thermodynamics, it is a critical mistake to have done so.

Properly including geothermal flux in a revised model not only accurately reproduces Earth's surface temperature but also provides a simple technique for measuring local, land-based, geothermal heat flow from space. The technique may also be useful in the study of global heat flow in ocean currents. This is illustrated by mapping excess heat flow in the Norwegian Sea attributable to the North Atlantic current (note the following image).

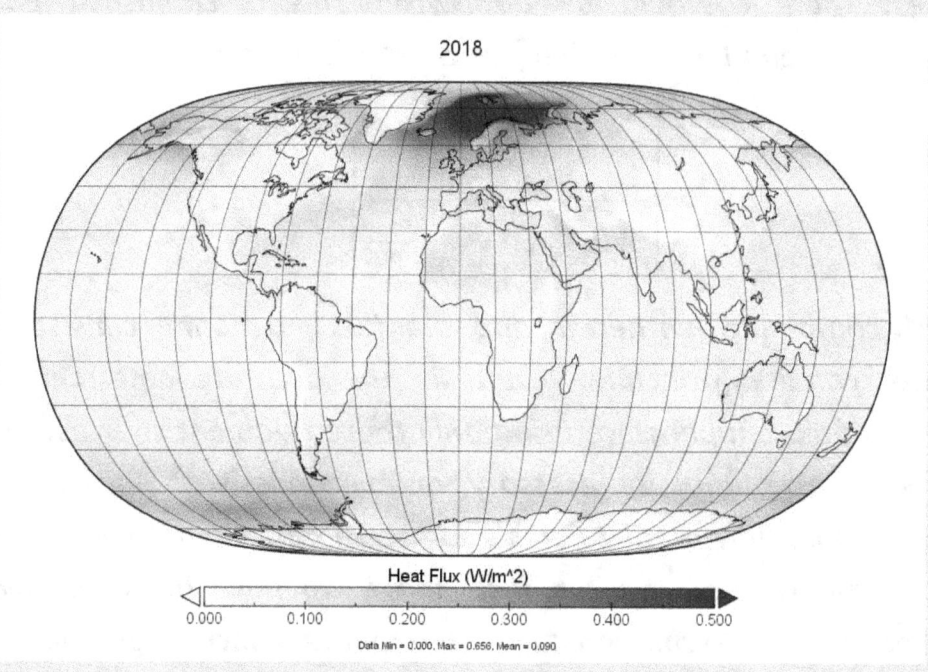

Figure 1 - 2018 global residual heat flow.

You would be hard-pressed to find someone on this planet that had not been exposed to a phrase similar to Nasa Science Collaborator, Qiancheng

Ma's [1], "Without naturally occurring greenhouse gases, Earth's average temperature would be near 0°F (or -18°C) instead of the much warmer [reality of] 59°F (15°C)." This statement is supported by the underlying relationship between an object's surface temperature and the rate at which energy is emitted from it in the form of electromagnetic radiation.

This relationship, credited to Josef Stefan and Ludwig Boltzmann [2], is stated

$$\Phi = \varepsilon \cdot \sigma \cdot T^4 \, , \qquad\qquad 1.1$$

where Φ is radiant flux, σ is the Stefan-Boltzmann constant $(5.67 \cdot 10^{-8} \; W/m^2K)$, ε represents a parameter known as the emissivity, and T is temperature. Solving this for T, we get

$$T = \left(\frac{\Phi}{\varepsilon \cdot \sigma}\right)^{0.25} = 64.8 \cdot \left(\frac{\Phi}{\varepsilon}\right)^{0.25} . \qquad 1.2$$

Earth's mean absorbed incident solar flux is approximately $240W/m^2$ [3] Conservation of energy requires that this same magnitude of flux must be radiated back into space when in equilibrium. For unit emissivity, this gives

$$T = 64.8 \cdot 240^{0.25} = 255K = -18C = -0.5F. \qquad 1.3$$

The best estimate of the current mean temperature of Earth's surface, as measured by numerous ground and satellite-based sensors, is $288K$ [4]. This Indicates that the Earth's surface is anomalously $33K$ warmer than it would be without some sort of modifying effect. Even if researchers included the heat from Earth's core, $0.065W/m^2$ [5] (as measured on land), the computed temperature appears essentially unchanged:

$$T = 64.8 \cdot 240.065^{0.25} = 255K = -18C = -0.5F. \qquad 1.4$$

This is the basis of the magnitude of the Greenhouse Effect which postulates that certain molecules in Earth's atmosphere, called Greenhouse Gasses, absorb and reradiate infrared radiation from Earth's surface [3]. Some of that reradiation is directed back to the surface where it is reabsorbed. This cyclic process is credited with maintaining the surface temperature at its elevated value. This effect is captured in Equation 1.2 as the emissivity parameter.

Using the observed *288K* value for Earth's mean surface temperature, we can compute the emissivity as

$$\varepsilon = 240 \cdot \left[\frac{64.8}{T}\right]^4 = 240 \cdot \left[\frac{64.8}{288}\right]^4 = 0.615 . \qquad \text{1.5}$$

However, most of Earth's surface has a measured emissivity of approximately *1.0* [6]. Therefore, the value in Equation 1.5 is anomalously low.

So, let's ask the question, "What would the surface temperature of the Earth be if there was no sun?" Using the Stefan-Boltzmann equation, unit emissivity, and only the heat flux from Earth's core we get

$$T = 64.8 \cdot 0.065^{0.25} = 33K. \qquad \text{1.6}$$

The result of this calculation being *exactly* the *33K* deficit surmised above is indication that there is something missing in the prevailing analysis. The possibility suggested by this result is that combining the incoming flux of two or more heat sources, q_i, in a steady state system cooled only by emission of radiation may not be according to the apparently obvious technique used in Equation 1.4, i.e.,

$$T = 64.8 \cdot \left(\sum_i q_i\right)^{0.25} . \qquad \text{1.7}$$

But, instead, via the technique suggested by the result in Equations 1.6 above,

$$T = 64.8 \cdot \sum_i (q_i^{0.25}).$$
1.8

This possibility motivates us to examine a more detailed model of the system at hand.

Figure 2 provides a graphical view of a standard, one-dimensional model of a planet's energy budget. The reader should view it as a column of 'dirt' extending some depth, L, into the planet. In this, Q_s is incident solar flux, Q_g is heat flux from the planet's interior to the surface, and Q_r represents cooling radiation emanating from the planet due to its surface temperature. The temperature at any time, t, and depth, y, is described by the function $T(y,t)$. The problem before us is to accurately determine that temperature function.

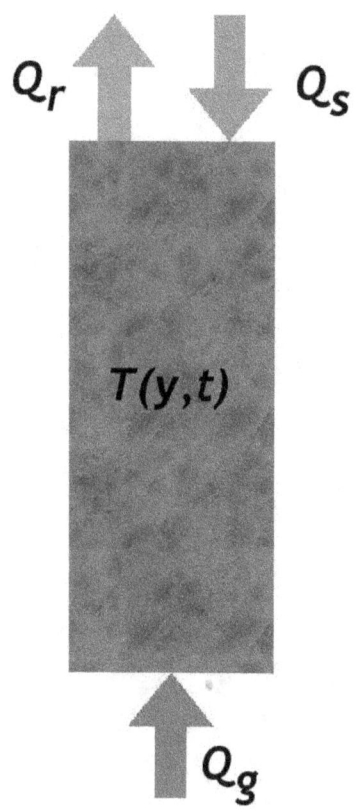

Figure 2 – A conceptual thermal column.

This is a type of problem covered in detail in standard graduate and advanced undergraduate programs in university departments of physics and mathematics world-wide. Mathematically, Figure 2 describes a *partial differential equation (PDE)* for which the Q values serve as boundary conditions.

The governing equation for our thermal column is known as the Heat Equation [7], which is a mathematical statement of four notable scientific observations, a few of which are alluded to in the chapter introduction:

- **First Law of Thermodynamics**

  In the study of our universe, we find that all detectable 'things' (objects, fields, collections of these, etc.) have the capacity to do work on other 'things'. By "doing work", we mean that one 'thing' transfers some of its capacity to another 'thing' or vice versa. The magnitude of this capacity is defined as energy and it cannot be created or destroyed. That is, energy is conserved. However, it can change form (potential, kinetic, electromagnetic, etc.). This concept of energy conservation is known as the First Law of Thermodynamics. Mathematically: $dE/dt = 0$.

- **Internal Energy**

  Macroscopic materials (dirt, glass, brick, air, etc.) have a property called *internal energy* whose magnitude is directly proportional to the material's mean temperature.

  Mathematically: $dU = c \cdot dT$ (where dU represents change in internal energy, dT means change in temperature, and c is the constant of proportionality).

- **Fourier Heat Equation**

  Internal energy can be transferred between adjacent regions inside a material. The rate at which this transfer takes place is called *heat* and its magnitude is directly proportional to the temperature difference between the adjacent regions.

Mathematically: $q = \eta \cdot dT/dy$ (where q is heat, y is a position within the material, and η is the constant of proportionality).

- **Stefan-Boltzmann Law**

  The phenomenon of a "red hot poker" demonstrates that internal energy can change form to electromagnetic energy (light). When this occurs at the surface of a substance, this electromagnetic energy will radiate away. Scientific observations show that the rate of this change is proportional to the fourth power of temperature.

  Mathematically: $\varphi = \varepsilon\sigma T^4$ (where σ is the constant of proportionality and ε is emissivity). Since Earth's surface area is primarily comprised of materials having an emissivity nearly equal to one, we will assume unit emissivity in this book.

These lead to the (in this case, one-dimensional) heat equation

$$\frac{\partial T(y,t)}{\partial t} - \kappa\frac{\partial^2 T(y,t)}{\partial y^2} = 0 , \qquad \text{1.9}$$

where κ is known as the *thermal diffusivity* of the material. In terms of the proportionality constants discussed and the density of the material, ρ,

$$\kappa = \frac{\eta}{\rho \cdot c} . \qquad \text{1.10}$$

The solution of this must adhere to the model of Figure 2. That is, heat flux from earth's core passes through the bottom of the column and solar flux enters the column from the top. Finally, the column top, owing to its temperature, must radiate energy to space according to the Stefan-Boltzmann law. These are our boundary conditions.

Both the solar flux and the planet's internal heat flux can be time dependent. Incoming solar flux varies in a periodic manner due to planet rotation, periodic changes in the planet's orbit, and the precession of its rotational axis. Herein, internal heat flux will be assumed largely time independent, which is consistent with observations. We will also arrange the depth of the column deep enough that solar variation effects will damp out before reaching that depth.

Solution to 1.9 is accomplished via the standard separation of variables technique:

$$T(y,t) = \tau(t) \cdot Y(y), \qquad\qquad 1.11$$

$$\frac{1}{\tau(t)} \frac{\partial \tau(t)}{\partial t} = \frac{\kappa}{Y(y)} \frac{\partial^2 Y(y)}{\partial y^2} = -i \cdot \lambda, \qquad\qquad 1.12$$

$$\tau(t) = e^{-i \cdot \lambda \cdot t}, \qquad\qquad 1.13$$

$$Y(y) = e^{(i-1) \cdot y \cdot \sqrt{\lambda/2\kappa}}, \qquad\qquad 1.14$$

$$\tau(t) \cdot Y(y) = e^{-i \cdot \lambda \cdot t + (i-1) \cdot y \cdot \sqrt{\lambda/2\kappa}}, \qquad\qquad 1.15$$

The imaginary unit is explicit in the exponentials since the incident solar flux is assumed periodic for the time scale of interest ($\pm 10^6$ years).

It is useful to deal with the $\lambda = 0$ condition as a special case. Equation 1.15 implies that, in that case, the solution is a constant. However, Equations 1.9 and 1.12 identify this condition as the steady-state (time-independent) solution. It does not, generally, indicate a fixed temperature throughout the thermal column. Temperature can still vary as a function of position provided it is consistent with Equation 1.9.

Therefore, the $\lambda = 0$ condition is restated as,

$$\frac{\partial T'(y)}{\partial y} = \varrho , \qquad\qquad 1.16$$

where ϱ is a constant. This is trivially solved as

$$T'(y) = \varrho \cdot y + \beta , \qquad\qquad 1.17$$

where β is a constant.

Since, in Equation 1.15, we are dealing with the homogeneous equation (no sources), the principle of superposition applies [7] (the superposition principle is generally stated as *"if there exist two or more solutions to a linear system then the weighted sum of those solutions is also a solution."*). Since λ in Equation 1.12 has no restriction on value, we can assume multiple possible values. If we label each possible value with index, n, we can use the superposition principle to write

$$T(y,t) = \sum_n \alpha_n \cdot e^{-i \cdot \lambda_n \cdot t + (i-1) \cdot y \cdot \sqrt{\lambda_n / 2\kappa}} \ (\lambda_n \neq 0), \qquad 1.18$$

where we specifically limit the solution to the time-dependent regime with $\lambda_n \neq 0$. The comprehensive solution, then, is

$$T(y,t) = \beta + \varrho \cdot y + \sum_n \alpha_n \cdot e^{-i \cdot \lambda_n \cdot t + (i-1) \cdot y \cdot \sqrt{\lambda_n / 2\kappa}} \qquad 1.19$$

Of useful importance here is the unmistakable superposition property of the solution. That is, the time-dependent and time-independent solutions can be determined separately, and those solutions added together in order to find the comprehensive behavior of the system. This is important for multiple reasons. The first is that contemporary analysis relies on the mean solar flux to compute the expected mean surface temperature. This analysis validates that technique. The second is that, once a steady-state solution is found (by whatever means), the time-dependent solution can be determined (again, by any means) and then simply added to the steady-state one in order to find the most general system behavior.

In order to resolve the question about what technique should be used to properly combine multiple heat sources in a **steady-state**, unit emissivity scenario, we will focus only on Equation 1.17 and ignore the time-dependence in Equation 1.19.

In steady state, the first law of thermodynamics instructs that the total outgoing flux **must** equal the total incoming flux. This makes the steady state problem almost trivial. If we define Q_s and Q_g in Figure 2, then we are also defining $Q_r = Q_s + Q_g$. This makes all boundary conditions of the *Neumann* variety. That is, the change of energy at the boundaries as a function of time,

$$\frac{dE}{dt} = C_p \frac{dT}{dt} = Q \, , \qquad 1.20$$

(C_p is the specific heat of the material in our thermal column, i.e., dirt) are specified as part of the problem statement.

So, what we have in steady state, is a manifestly linear system described by Equation 1.17 where all boundary conditions are of the *Neumann* variety. As such, this system meets all the requirements necessary to allow the use of the superposition principle [7].

A more transparent statement of the superposition principle is *"two or more excitations of a linear system will elicit a net response from the system equal to the sum of the responses that would have been caused by each individual excitation separately."* Specifically applied to the heat equation of interest, the superposition principle indicates that we can solve the equation for each boundary condition separately and then add those solutions together to assemble the complete mathematical model. Graphically,

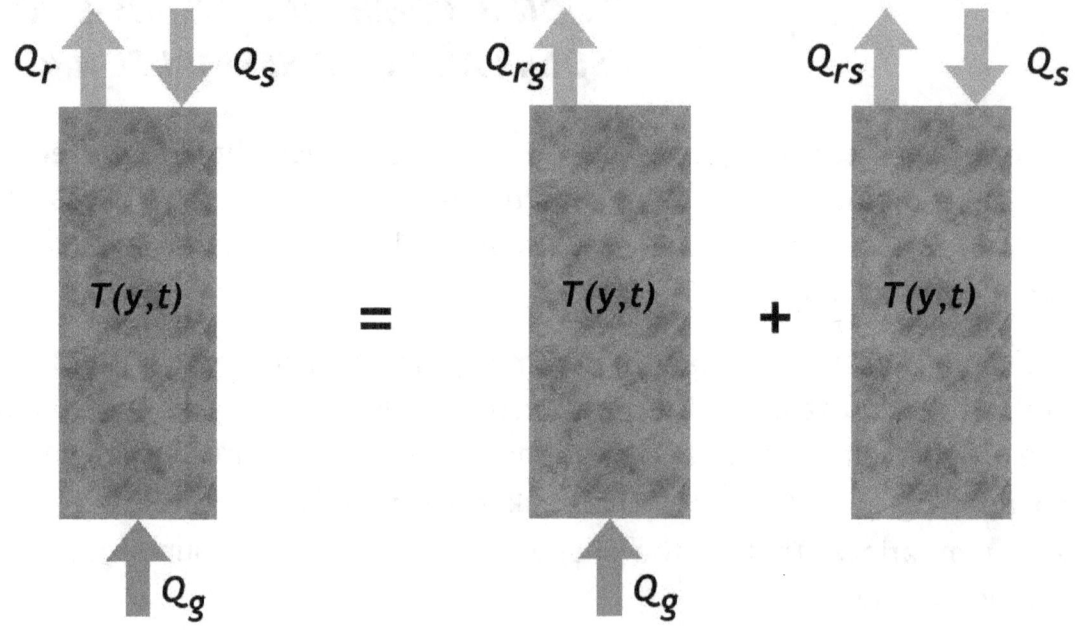

Figure 3 - Thermal column equivalency using superposition theorem.

Since, under a steady-state assumption, the superposition principle can be justifiably applied to Equation 1.17, the general solution to the problem is the sum of the solutions where solar and geothermal flux are treated independent of each other. Further, since the energy leaving the thermal column at the top surface does so via radiation, the Stefan-Boltzmann equation can be used to determine the temperature necessary in order to accomplish that radiation.

It is known that ground temperature increases as one probes to greater depths. This is known as the geothermal gradient. Globally, the mean increase in temperature is about *27C (or, equivalently, K) per km* of depth. Fourier's Law and the Second Law of Thermodynamics clearly indicate that heat flows from warmer regions to colder regions and not vice versa. Therefore, in steady-state, solar heating of the surface does not penetrate to lower depths. It cannot, since that would violate the Second Law. So, in the steady-state view, heat flow is *always* directed toward the surface of the planet.

Therefore, solving the solar flux by itself provides

$$T_S(y) = 64.8 \cdot 240^{0.25} = 255K \,, \qquad 1.21$$

where the temperature of the thermal column is uniform throughout as required by the previous discussion, i.e., steady state without downward heat flow.

Considering the geothermal flux by itself gives,

$$T_G(y) = 0.027 \cdot y + 64.8 \cdot 0.065^{0.25} = 0.027y + 32.7 \,, \qquad 1.22$$

where the *0.027* value represents the average *27K/km* temperature change with depth found worldwide [8]. The *0.065W/m²* is the mean geothermal flux as measured on land.

As prescribed by the superposition principle,

$$T(y) = T_S(y) + T_G(y) = 0.027y + 287.7 \,. \qquad 1.23$$

This completes the detailed mathematical proof that the proper method of combining multiple heat sources in a **steady-state**, linear system cooled by Stefan-Boltzmann-induced radiation is as shown in Equation 1.8 and not the contemporary technique of Equation 1.7. Note that it is overkill to resort to the heat equation when Fourier thermal conductivity would suffice. However, completeness won out.

This result conclusively shows that the current and ubiquitous thought that Earth's surface temperature would be substantially colder, specifically *33K* colder, if not for the atmospheric Greenhouse Effect, is incorrect. Earth's surface temperature is fully explained by the construction in this section. In fact, it is revealed here that the Greenhouse Effect, itself, is the result of erroneously combining the solar and geothermal fluxes in a manner inconsistent with standard, applicable PDE solution techniques; i.e., the superposition principle.

The reader should understand that the superposition principle is not simply a useful technique for solving partial differential equations. In this case, the linearity of the system that allows the use of the superposition principle is the first law of thermodynamics, i.e., conservation of energy. No matter what the source of energy is at some position in our thermal column, that energy will be equivalent to the temperature. If there are two distinct energy sources, conservation of energy requires that the resulting magnitude of their combination be just their sum. It cannot be less without destroying energy and it cannot be more without creating energy. Both possibilities would violate the First Law.

Consequently, a solution of the steady-state heat equation subject to Dirichlet and/or *Neumann* boundary conditions that does not obey the superposition principle manifestly violates the first law of thermodynamics. Therefore, to conclude that the Greenhouse Effect theory is responsible for *33K* of Earth's mean surface temperature is to manifestly violate the First Law in that it relies on combining energy flux as given by Equation 1.7 which does not obey the superposition principle.

As discussed in the last section, time dependence of the incoming solar flux is largely due to Earth's rotation, the angle its rotation axis makes with the ecliptic, and changes in its orbit about the sun. The first task in fleshing out the time-dependent part of the thermal column model is to characterize the solar flux variation. In this section we will focus on present day variations. In the next chapter we will tackle the paleoclimate problem.

Celestial Coordinate System

The celestial plane is defined by Earth's orbit about the sun. This orbit is elliptical in nature and is accurately modeled according to the equation [9]

$$R(\chi) = \frac{a \cdot (1 - e^2)}{1 - e \cdot \cos(\chi)}. \qquad\qquad 1.24$$

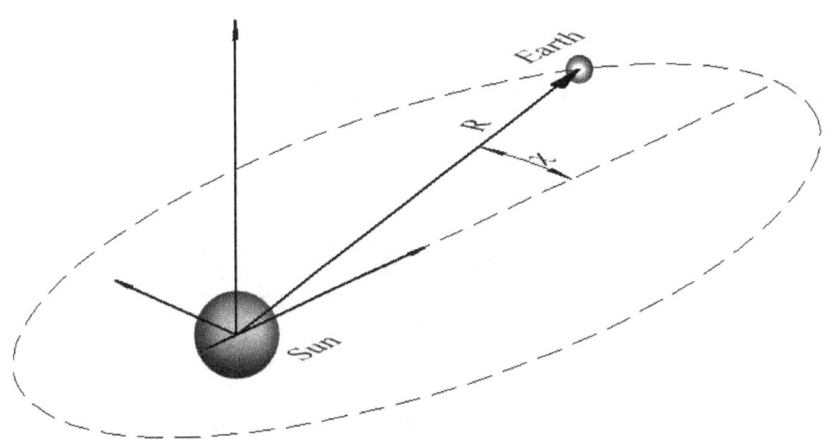

Figure 4 - Celestial plane and coordinate system showing Earth's orbit. The ellipse here is exaggerated compared to Earth's nearly circular orbit.

In this, a is the semimajor axis of the ellipse (half the long 'diameter') and e is the eccentricity. The current values [10] of these parameters are *149.6 million kilometers* and *0.0167,* respectively. Adding any constant to χ simply rotates (precesses) the orbit in space and has no material effect on the analysis.

Kepler's third law determines orbital period [9] as

$$T = 2 \cdot \pi \sqrt{\frac{a^3}{\mu}} ,$$
1.25

where μ is the gravitational parameter

$$\begin{aligned} \mu &= G \cdot M \\ &= (6.67408 \times 10^{-20}) \cdot (1.989 \times 10^{30}) . \\ &= 1.327 \times 10^{11} \frac{km^3}{s^2} \end{aligned}$$
1.26

G is the gravitational constant and M is the mass of the sun. This gives Earth's orbital period as

$$\begin{aligned} T &= 2 \cdot \pi \cdot \sqrt{\frac{(149.6 \times 10^6 km)^3}{1.327 \times 10^{11} \frac{km^3}{s^2}}} \\ &= 3.155 \times 10^7 s . \end{aligned}$$
1.27

In order to obey the fundamental law of physics that requires conservation of angular momentum, Earth's angular velocity increases when closer to the sun and slows down when it is farther away. So, to find its orbital position with respect to time requires a bit of math:

$$R(X) \cdot \dot{X} = \frac{a \cdot 2 \cdot \pi}{T} \qquad\qquad 1.28$$

$$\frac{T}{2 \cdot \pi} \cdot \frac{(1 - e^2)}{1 - e \cdot \cos(X)} dX = dt \qquad\qquad 1.29$$

$$\frac{T}{2 \cdot \pi} \cdot \int \frac{(1 - e^2)}{1 - e \cdot \cos(X)} dX = t \qquad\qquad 1.30$$

$$-\frac{T \cdot \sqrt{1 - e^2}}{\pi} \cdot \arctan\left[\frac{\tan\left(\frac{X}{2}\right) \cdot \sqrt{1 - e^2}}{e - 1}\right] = t \qquad\qquad 1.31$$

$$X(t) = -2 \cdot \arctan\left[\frac{\tan\left(\frac{\pi \cdot t}{T \cdot \sqrt{1 - e^2}}\right) \cdot (e - 1)}{\sqrt{1 - e^2}}\right] \qquad\qquad 1.32$$

$$X(t) = 2 \cdot \arctan[0.983 \cdot \tan(9.959 \times 10^{-8} \cdot t)] \qquad\qquad 1.33$$

where t is given in seconds. (I've avoided the use of astronomer's *eccentric* and *mean anomalies* for the sake of familiarity.)

Time, here, will be referenced to the Earth's rotation with respect to distant stars. One rotation, in this frame of reference, takes 23 hours, 56 minutes, and four seconds *(86,164 s)*. Therefore, using ε as Earth's angle of rotation in radians, and ε_0 as the beginning rotation angle at aphelion,

$$X(\epsilon) = 2 \cdot \arctan[0.983 \cdot \tan(1.366 \times 10^{-3} \cdot (\epsilon - \epsilon_0)]. \qquad\qquad 1.34$$

Ecliptic Coordinate System

The ecliptic plane (generally referred to as simply the ecliptic) is defined by an infinite plane coincident with Earth's equator. Earth's axis of rotation is perpendicular to this plane. A point on Earth's surface will be designated by its spherical polar coordinates. In standard notion, ϕ will be the polar angle measured from the north extension of the rotation axis, z'. In the more familiar latitude/longitude measurement, the polar angle is 90 degrees plus the south latitude angle or 90 degrees minus the north latitude angle. Similarly, θ will be the azimuthal angle measured from zero degrees latitude (the prime meridian) increasing going east (the direction of earth's rotation). This coordinate system is identified by the prime accents on the direction vectors.

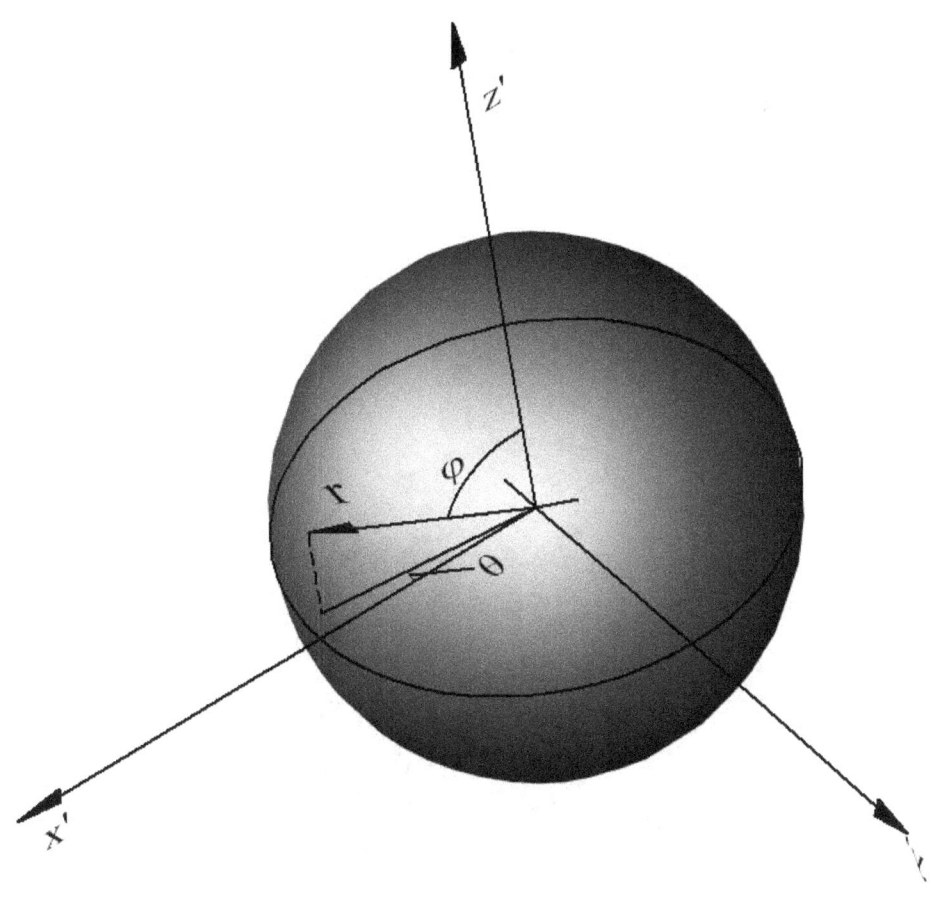

Figure 5 - Ecliptic coordinate system. Earth's rotation axis is labeled Z'.
The foreshortened circle inside the sphere represents the ecliptic plane.

Coordinate Transformation

The planet's rotation axis *(z')* is tilted with respect to a vector normal to the celestial plane *(z)*. The celestial plane is the plane in which the Earth orbits the sun. This tilt is known as the obliquity and it is currently *0.408 radians (23.4°)* [2].

The projection of that tilt onto the celestial plane presently aligns with the Θ = *3.302 radians (189.191°)* in Figure 4. This is known as the *precession angle* which slowly changes over time.

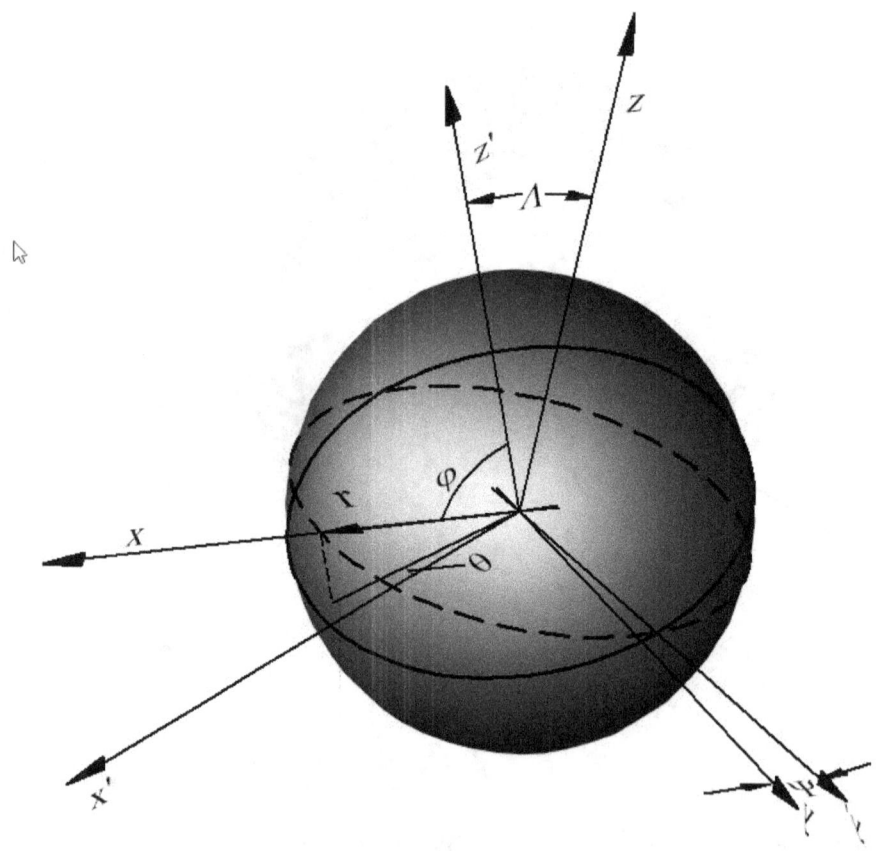

Figure 6 - Coordinate transformation.
Primed coordinates are ecliptic. Unprimed are celestial.
Solid foreshortened circle represents the ecliptic plane.
Dashed foreshortened circle represents the celestial plane.

Obliquity is important in defining seasons. When the Earth is canted toward the sun, it is summer in the northern hemisphere and winter in the southern. When canted away, the seasons are reversed.

By inspection of Figure 6, coordinate translation from ecliptic to celestial systems involves a rotation around the *z-axis* of -Ψ radians followed by a rotation about the *y-axis* of -Λ radians. Or, in matrix form,

$$R_z(\Psi) = \begin{bmatrix} \cos(\Psi) & \sin(\Psi) & 0 \\ -\sin(\Psi) & \cos(\Psi) & 0 \\ 0 & 0 & 1 \end{bmatrix} \qquad 1.35$$

$$R_y(\Lambda) = \begin{bmatrix} \cos(\Lambda) & 0 & -\sin(\Lambda) \\ 0 & 1 & 0 \\ \sin(\Lambda) & 0 & \cos(\Lambda) \end{bmatrix} \qquad 1.36$$

$$\begin{aligned} R_{tot}(\Psi,\Lambda) &= R_z(\Psi) \cdot R_y(\Lambda) \\ &= \begin{bmatrix} \cos(\Psi)\cdot\cos(\Lambda) & \cos(\Lambda)\cdot\sin(\Psi) & -\sin(\Lambda) \\ -\sin(\Psi) & \cos(\Psi) & 0 \\ \cos(\Psi)\cdot\sin(\Lambda) & \sin(\Psi)\cdot\sin(\Lambda) & \cos(\Lambda) \end{bmatrix} \end{aligned} \qquad 1.37$$

$$\begin{aligned} R_{tot}&(-3.302,-0.408) \\ &= \begin{bmatrix} -0.906 & -0.147 & -0.397 \\ 0.16 & 0.987 & 0 \\ -0.392 & -0.063 & 0.918 \end{bmatrix} \end{aligned} \qquad 1.38$$

using the current obliquity and precession angle of Earth's rotation axis.

Solar Flux Reduction Due to Geometrical Aspect

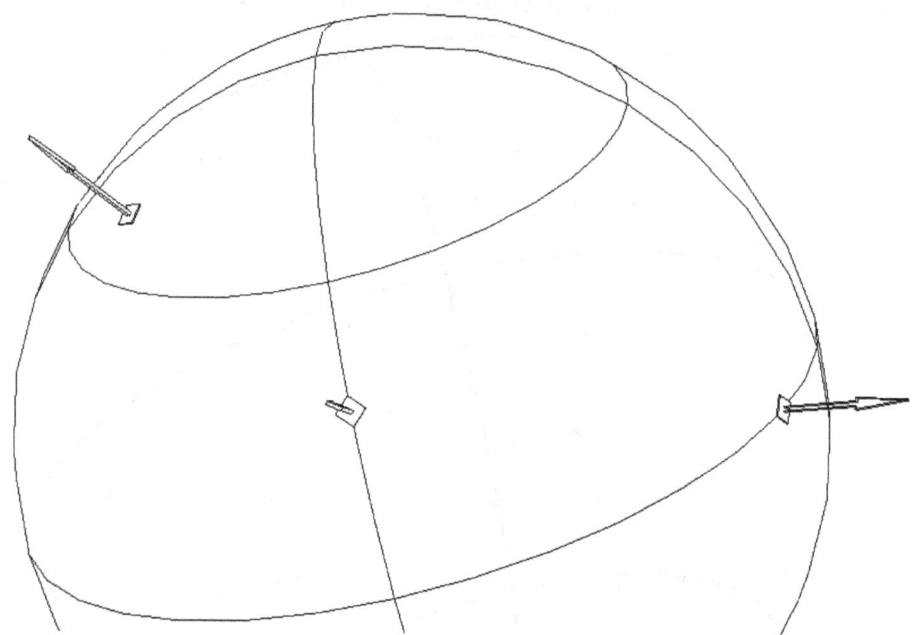

Figure 7 - Three equal-area regions on the surface of the planet, as seen from the sun. The area presented to the sun is reduced by the cosine of the angle the region is turned away.

With reference to Figure 7, incident solar power *(W)* striking a point on the planet surface is the product of the solar flux *(W/m²)* and the small area *(m²)* surrounding that point. The size of that area is reduced by its aspect to the sun as the cosine of the angle between the solar ray and the surface normal vector at the point of interest.

This reduction is determined as the vector dot product of the incident solar ray direction vector and unit surface normal of the point of interest. The incident solar ray direction vector is just *R* in Figure 4 divided by its magnitude. This is given in the celestial coordinate system. Note that the planet-sun separation distance allows the assumption of parallel rays. This means only the *x-y* components of the incoming solar rays are germane in performing the dot product.

The unit normal of the point of interest is given by the polar and azimuthal angles *(φ and θ, respectively)* shown in Figure 5 and 4. In vector notation, the unit normal to a point, *P'*, on the Earth's surface would be written (using the prime accent on *P* to indicate the ecliptic coordinate system),

$$\hat{P}'(\varphi, \theta', \epsilon) = \begin{bmatrix} \sin(\varphi) \cdot \cos(\theta' + \epsilon) \\ \sin(\varphi) \cdot \sin(\theta' + \epsilon) \\ \cos(\varphi) \end{bmatrix}$$
$$= \begin{bmatrix} \sin(\varphi)\cos(\theta')\cos(\epsilon) - \sin(\varphi)\sin(\theta')\sin(\epsilon) \\ \sin(\varphi)\sin(\theta')\cos(\epsilon) + \sin(\varphi)\cos(\theta')\sin(\epsilon) \\ \cos(\varphi) \end{bmatrix}.$$

1.39

In this,

$$\theta' = \theta - \epsilon_0 ,$$

1.40

where ϵ_0 is the beginning rotation angle of the Earth at aphelion, as discussed in Equation 1.34.

Conversion of $P'(\varphi, \theta)$ to celestial coordinates is necessary to perform the intended dot product. This is accomplished by matrix multiplication of Equations 1.38 and 1.39,

$$\hat{P}(\varphi, \theta') = R_{tot}(-3.302, -0.408) \cdot P'(\varphi, \theta')$$
$$= \begin{bmatrix} -0.906 & -0.147 & -0.397 \\ 0.16 & 0.987 & 0 \\ -0.392 & -0.063 & 0.918 \end{bmatrix}$$
$$\cdot \begin{bmatrix} \sin(\varphi)\cos(\theta')\cos(\epsilon) - \sin(\varphi)\sin(\theta')\sin(\epsilon) \\ \sin(\varphi)\sin(\theta')\cos(\epsilon) + \sin(\varphi)\cos(\theta')\sin(\epsilon) \\ \cos(\varphi) \end{bmatrix},$$

1.41

$$P_x(\,(\varphi,\theta',\epsilon) = -0.397 \cdot \cos(\varphi)$$
$$- (0.906 \cdot \cos(\theta') + 0.147 \cdot \sin(\theta')) \cdot \sin(\varphi) \cdot \cos(\epsilon) \qquad 1.42$$
$$+ (0.906 \cdot \sin(\theta') - 0.147 \cdot \cos(\theta')) \cdot \sin(\varphi) \cdot \sin(\epsilon)$$

$$P_y(\,(\varphi,\theta',\epsilon) =$$
$$-(0.987 \cdot \cos(\theta') + 0.16 \cdot \sin(\theta')) \cdot \sin(\varphi) \cdot \sin(\epsilon) \qquad 1.43$$
$$+ (-0.987 \cdot \sin(\theta') + 0.16 \cdot \cos(\theta')) \cdot \sin(\varphi) \cdot \cos(\epsilon)$$

where the x and y vector components are written explicitly only for clarity and space limitations. The z component is not shown since it will not participate in the dot product.

Using Figure 4 and Equation 1.34, the solar flux direction vector is given by

$$\widehat{\Phi}(\epsilon) = \begin{bmatrix} \cos[X(\epsilon)] \\ \sin[X(\epsilon)] \\ 0 \end{bmatrix}. \qquad 1.44$$

This gives the dot product we are seeking as

$$\widehat{\Phi}(\epsilon) \cdot \widehat{P}(\varphi,\theta')$$
$$= \cos[X(\epsilon)] \cdot P_x(\,(\varphi,\theta',\epsilon) \qquad 1.45$$
$$+ \sin[X(\epsilon)] \cdot P_y(\,(\varphi,\theta',\epsilon).$$

Note that only negative values of this dot product are valid since positive values indicate that the two vectors are pointing in the same direction. That is, the body of the planet is between the surface element and the sun. So, positive values need to be set to zero.

Solar Flux Reduction Due to Sun-Planet Separation Distance

As shown graphically in Figure 4, due to the elliptical nature of a planet's orbit, the position in the orbit is also of import when determining incoming solar flux. If we imagine a spherical shell with radius $\acute{R}$ about the sun, then the mean solar flux through that shell will be

$$|\Phi| = \frac{P^{\odot}}{4 \cdot \pi \cdot \acute{R}^2}. \qquad 1.46$$

where $P^{\odot}$ is the total power output of the sun, $3.826 \times 10^{26} W$.

Solar Flux Reduction Due to Albedo

Albedo is a term that describes incident solar radiation that is reflected from a planet without being absorbed. This is due to reflectivity of the surface terrain or intervening cloud cover. Current measurements show that Earth's average albedo, α, is approximately *0.29* [11]. This modifies Equation 1.46 as,

$$
\begin{aligned}
|\Phi|' &= \frac{(1 - \alpha) \cdot P^{\odot}}{4 \cdot \pi \cdot \acute{R}^2} \\
&= \frac{0.71 \cdot P^{\odot}}{4 \cdot \pi \cdot \acute{R}^2} \\
&= \frac{2.18 \times 10^{25}}{\acute{R}^2} \frac{W}{m^2}.
\end{aligned}
\qquad 1.47
$$

Using Equation 1.24 and the current values of eccentricity and semi-major axis, this becomes

$$
\begin{aligned}
|\Phi(\epsilon)|' &= \frac{2.18 \times 10^{25} \cdot [1 - 0.0167 \cdot \cos(X(\epsilon))]^2}{[a \cdot (1 - e^2)]^2} \\
&= 967 \cdot [1 - 0.0167 \cdot \cos(X(\epsilon))]^2 \cdot \frac{W}{m^2}.
\end{aligned}
\qquad 1.48
$$

Solar Flux Incident on a Particular Location on Earth's Surface

Combining geometrical aspect, sun-planet separation distance, and albedo effects, the solar flux incident on a specific location on Earth's surface is given as

$$|\Phi(\epsilon, \varphi, \theta')|' = 967 \cdot \left[1 - 0.0167 \cdot \cos(X(\epsilon))\right]^2 \cdot$$
$$\min\left[\left[\cos\left[X(\epsilon)\right] \cdot P_x(\ (\varphi, \theta') + \ \sin[X(\epsilon)] \cdot P_y(\ (\varphi, \theta')\right], 0\right]$$

1.49

in W/m^2. Note that the "min" function serves to limit the geometrical aspect reduction to conditions where the planet surface element is angled toward the sun.

The mean solar flux incident on a specific location on Earth's surface is given by averaging Equation 1.49 over one orbit around the sun (*365.25* days),

$$\overline{\Phi(\varphi,\theta)} = \frac{1}{365.25 \cdot 2 \cdot \pi} \cdot \int_0^{365.25 \cdot 2 \cdot \pi} |\Phi(\epsilon,\varphi,\theta)|' d\epsilon. \qquad 1.50$$

where ε as given in Equation 1.5 is used.

Using the result of Section B, the relationship among the mean values of surface temperature, T, geothermal flux, q_g, and solar flux, Φ, for a specific point on Earth's surface is given by

$$T_{FS}(\varphi,\theta) = \left[\frac{\overline{\Phi(\varphi,\theta)}}{\sigma}\right]^{0.25} + \left[\frac{q_g}{\sigma}\right]^{0.25}. \qquad 1.51$$

If one knows any two of these, then the third can be determined by Equation 1.51.

This last statement essentially allows the use of a thermometer to determine the mean geothermal flux of any location on a planet. This should prove useful for satellite sensing. To that end, Equation 1.50 was evaluated on a one-degree by one-degree grid of Earth's surface.

Using the baseline average (Jan 1951 – Dec 1980) Berkeley Earth (BEST [12]) data for $T_{FS}((\varphi,\theta)$ and the calculations of Equation 1.50 in Equation 1.51, it is trivial to solve for q_g. The resulting dataset provides a geothermal flux map of the entire planet:

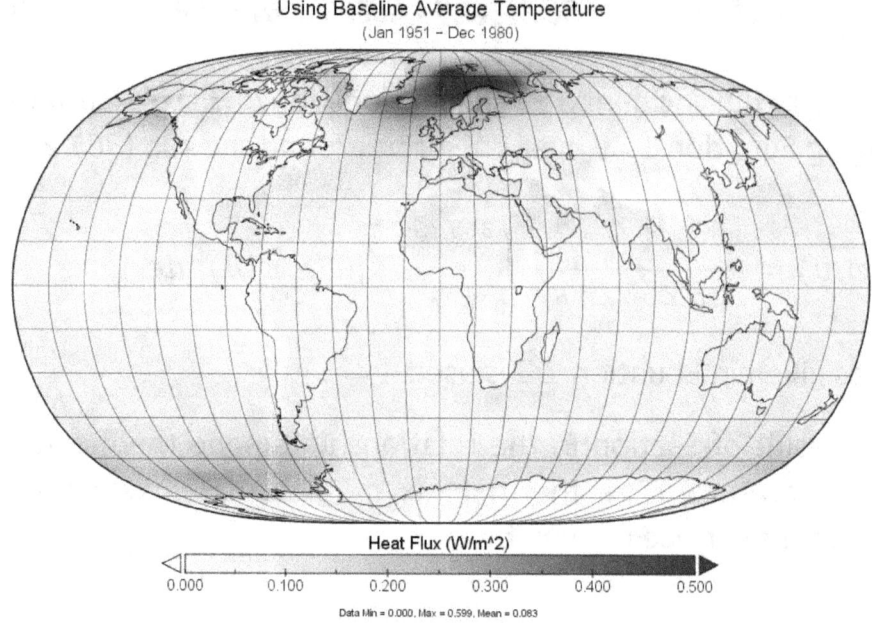

Using Baseline Average Temperature
(Jan 1951 – Dec 1980)

Heat Flux (W/m^2)

0.000   0.100   0.200   0.300   0.400   0.500

Data Min = 0.000, Max = 0.599, Mean = 0.083

Figure 8 – Baseline global geothermal heat flux.

It is impossible to ignore the prominent flux region centered in the Norwegian Sea. Even though this is a well-known, highly active volcanic area (Iceland itself has a significant volcanic eruption every three or four years), this *Norwegian Spot* is not, primarily, due to the volcanism. To deduce this, note that there are no other active volcanic regions represented to this extent on the plot (Hawaii, Yellowstone, etc.). Instead, this *Spot* appears mainly the result of the North Atlantic current which shuttles warm water into that region. In fact, this technique of acquiring radiative heat flux from temperature measurements should prove useful in analyzing the contribution of ocean currents in warming polar regions.

It should be noted that the *Spot* and other enhanced radiative flux regions seen are artifacts more than geothermal flux readings. The enhanced flux found from the computation obtain, primarily, from the warm water flowing into the region and not from local geothermal activity. Therefore, the term *residual flux* is probably more accurate than *geothermal flux* when referring to the q_g term in Equation 1.51 when dealing with ocean regions.

Using historical temperature data from Berkeley Earth, it is evident that this *Spot* is not a transient phenomenon. The following residual flux plots for 1850 and 2018 show the *Spot's* persistence.

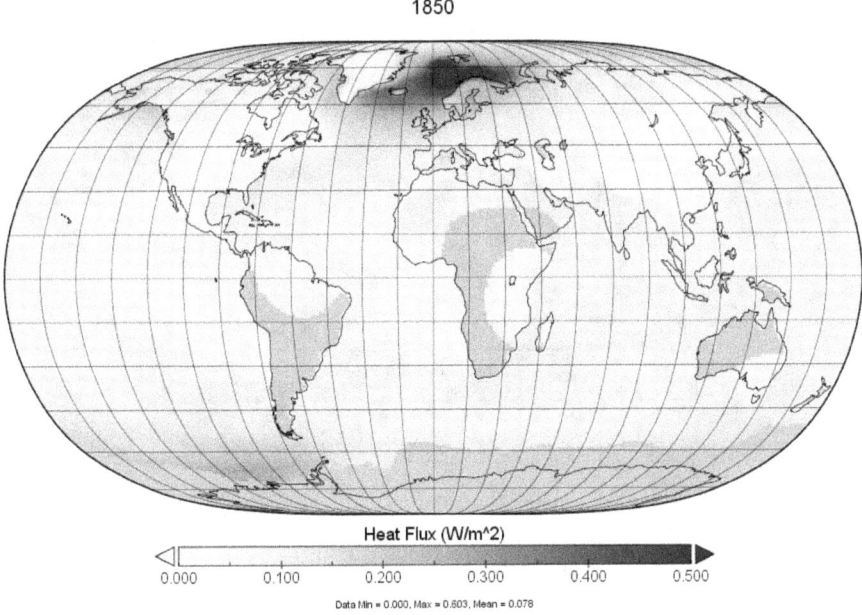

Figure 9 - 1850 global geothermal heat flux.
Some regions had no available data.

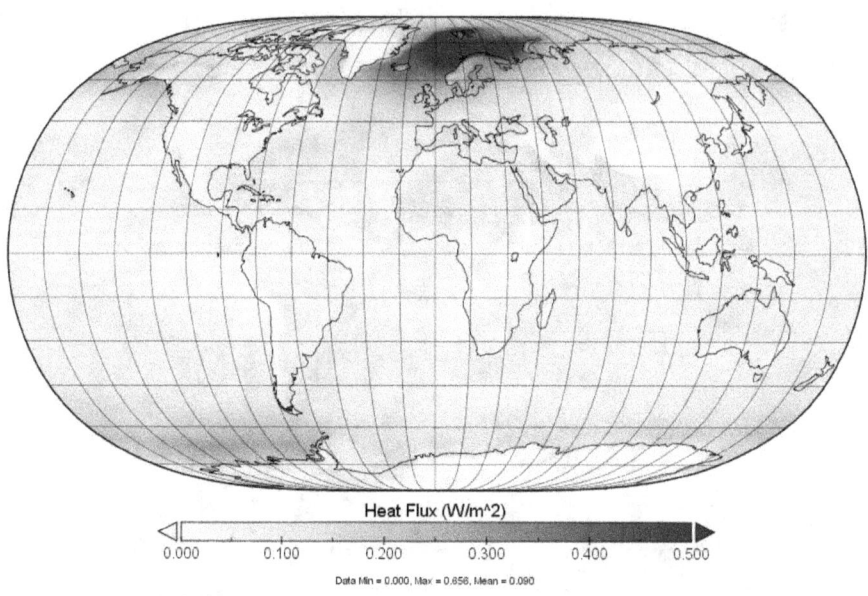

Heat Flux (W/m^2)

0.000   0.100   0.200   0.300   0.400   0.500

Data Min = 0.000, Max = 0.656, Mean = 0.090

Figure 10 - 2018 global geothermal heat flux.

Visually, the *Spot* in 2018 is more prominent than that shown in the baseline data. It is also more prominent than in 1850, but the missing temperature data in 1850 makes this less suggestive.

The *Spot,* based on the point of maximum computed residual flux within it, is centered in the vicinity of *72.6(±1.4)°N* latitude, *13.2(±1.9)°E* longitude. One can get a feel for the relationship between the *Spot* flux and the global mean temperature by plotting the anomaly of residual flux (multiplied by a factor of 10) alongside the global mean temperature anomaly:

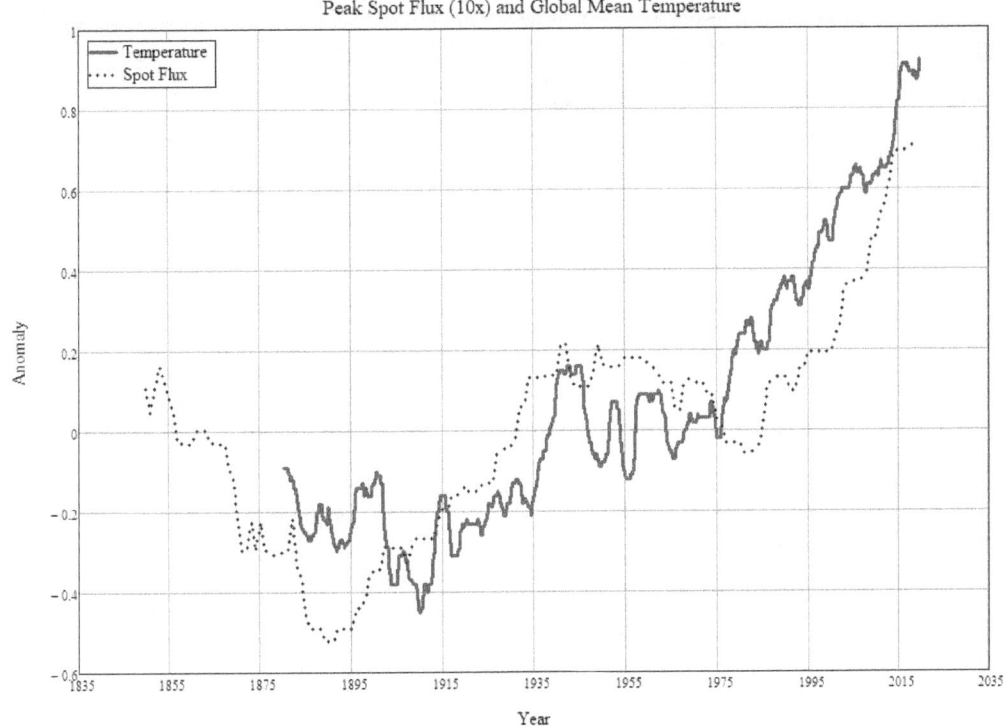

Figure 11 - Temperature and *Spot* flux anomalies (smoothed).

Finally, close examination of Figure 8 gives the impression of anomalously low geothermal flux in mountainous regions. This is presumably due to temperature measurements being made at altitude which can be compensated for using the typical temperature lapse rate [13] 0f 0.0065 C/m. This compensation is shown in Figure 12 for the Berkeley Earth baseline dataset.

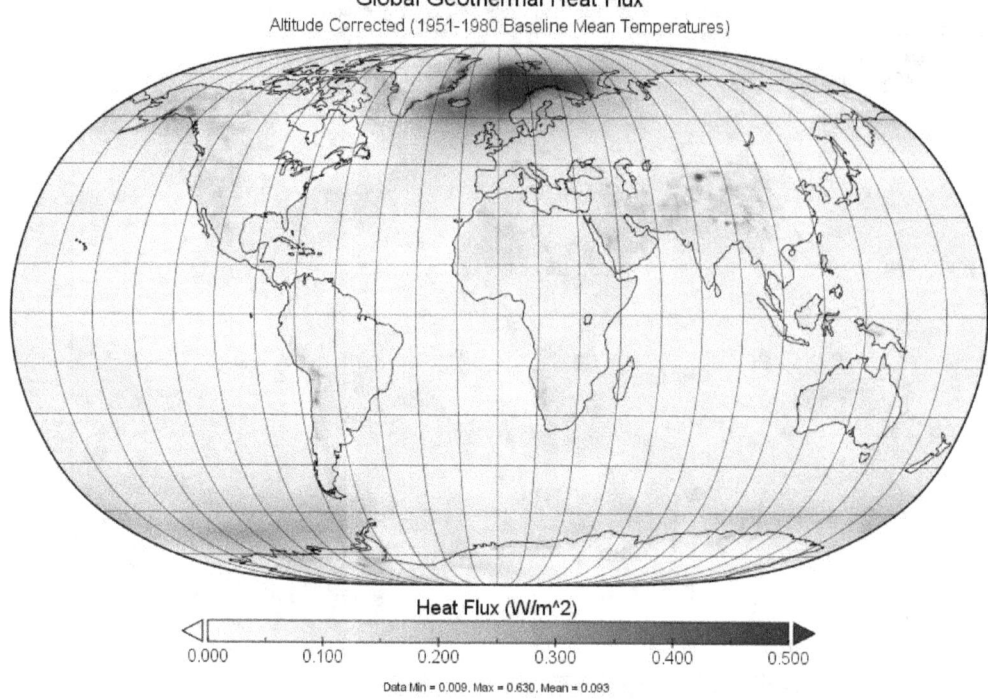

Figure 12 - Altitude-Corrected Geothermal Heat Flux.

Orbit Mean Summary

The orbit-mean residual flux solution draws a consistent parallel between geothermal flux and planet temperatures in excess of that attributable to solar flux alone. This is an expected result based on the mathematical construction of this and previous sections.

The trivial computation of global residual flux should serve as an important technique for analyzing ocean current effects and trends using widely available global temperature measurements.

Note that the construction of the geothermal mapping figures provides further evidence that the Greenhouse Effect is the result of a mathematical error, i.e., neglecting the superposition principle and violating the first law of thermodynamics, and is categorically not a physical phenomenon.

SECTION E. "ACTUAL" MEAN SURFACE TEMPERATURE

There have been numerous treatises written on the challenges encountered when attempting to measure the global mean temperature [14]. Ideally, we would like to have large number of identical temperature sensors spaced evenly around the globe at the same altitude. These would continuously acquire data for an extended time so that any periodic changes in climate would be captured and properly accounted for. Once the dataset was acquired, we could average them both locally and globally for comparison with subsequent data.

When researchers began to focus on the possibility of anomalous changes in global temperatures, this sort of dataset was not available. Using the large amount of data that was available worldwide, they picked a *30*-year period to use as the "baseline" dataset. The *30*-year period was chosen based on the US National Weather Service guidance that a three-decade period adequately defines "normal" or average temperature.

The datasets from 1951 through 1980 were selected as the baseline set since earlier datasets were not as extensive and later datasets might be contaminated by the greater extent of industrial activities. The dataset used was limited to land-based, surface temperatures as they are fixed locations easier to reliably monitor on an individual basis.

Even with such fixed locations, data can be subject to systematic error. For example, if a monitoring station is near a smelter, the temperature recorded may be elevated with respect to the actual natural temperature of that location.

As such, it was determined that temperature "anomalies" (temperature differences from the baseline) would be used to determine and report trends in local and global mean temperatures.

This is very important in the analysis on this chapter because it indicates the extent to which variations in geothermal heat flux affect the temperature data used to determine global mean surface temperature.

Recall that applying the Stefan-Boltzmann equation to the land-based mean geothermal heat flux ($0.065W/m^2$) fully explains the $33K$ temperature deficit that prompted the need for the Greenhouse Effect theory in the first place. That $0.065W/m^2$ value comes from *Pollack, et al.* [5]. As mentioned, this is the mean heat flow over continental crust. The related value over ocean crust is $0.101W/m^2$, making the global average $0.087W/m^2$.

So,

- the baseline temperature data was collected entirely from land-based stations over continental crust,
- the mean computed from the baseline temperature data is $33K$ greater than can be explained by solar flux alone, and
- the continental crust heat flux of $0.065W/m^2$ exactly reproduces that $33K$ temperature.

So, even though there exists a global mean geothermal heat flux of $0.087W/m^2$, the local heat flux dictates the local temperatures. This last indicates the legitimacy of the global residual heat flux maps of SECTION D.

Using the $0.087W/m^2$ value, the true global mean geothermal-sourced temperature is $35.2K$. Not the $33K$ found empirically while restricting data to heat flows above continental crust.

If we plot the computed mean residual flux for the entire dataset (1850-2018), we obtain

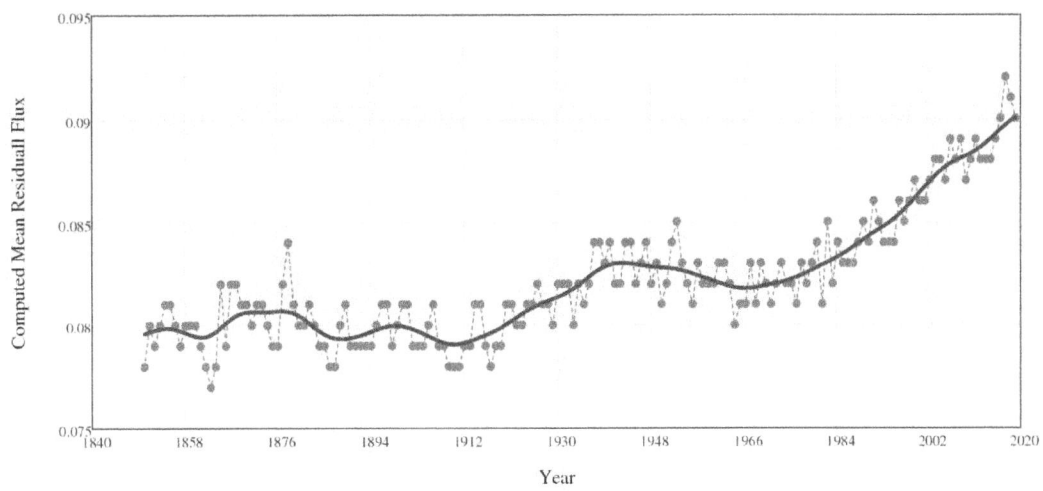

Figure 13 - Computed Annual Mean Residual Flux.

If we rely on the *0.087W/m²* value from *Pollack, et al.,* we can see that the data transitions from below that value to above. This appears to imply that the measured increase in annual mean global temperature anomalies might be due to geothermal flux changes. This, however, is overreaching. The technique for computing the residual flux assumes any temperature at variance with that due to a fixed solar flux is "residual". So, "residual" does not have to be "geothermal".

That said, it is notable that the residual flux seems to be confined to the region around the mean geothermal flux value provided by *Pollack, et al.* It suggests a possible geothermal oscillation being responsible for "fine structure" global mean temperature variations.

Chapter 2 - Executive Summary

This chapter exposes an error made in orbit calculations that vastly underestimate the effect of orbit eccentricity variations on insolation values. This error shrouds the extent to which Milankovitch's theories were entirely correct in attributing global climate changes to "orbital forcing".

The relationship between orbit eccentricity and the amount of solar energy striking Earth's surface (insolation) is considered "settled science". Mathematically, it is written

$$W_{am}(e) = \frac{S}{4} \cdot (1 - e^2)^{-\frac{1}{2}}.$$

However, despite the widespread agreement on this, the relationship is in error. This can be easily shown by examining two cases. The first would be a circular orbit where eccentricity is zero:

$$W_{am}(0) = 0.250 \cdot S.$$

The second being an eccentricity of, say 0.9 (almost parabolic):

$$W_{am}(0.9) = 0.574 \cdot S.$$

This Indicates that an orbit in which the planet is rarely in close proximity to the sun has over twice the insolation than it would in a circular orbit near its perihelion.

Since the circular orbit is the lowest energy state of all possible orbits, it is the closest possible mean distance to the sun. Yet, the equation above indicates that insolation is greater for orbits on which planets spend more time distant from the sun. That view defies common logic and this chapter will resolve it with the correct relationship

$$W_{am}(e) = \frac{S_0}{4} \cdot \sqrt{\frac{(1-e)^3}{(1+e)}} \text{ ,''}$$

where S_0 **is the solar constant of the planet's circular orbit about the sun defined by its angular momentum.**

The following is a graphical representation of temperature as a function of time for both relationships (dotted/blue for the flawed version and solid/red for the corrected one) by using the Stefan-Boltzmann equation.

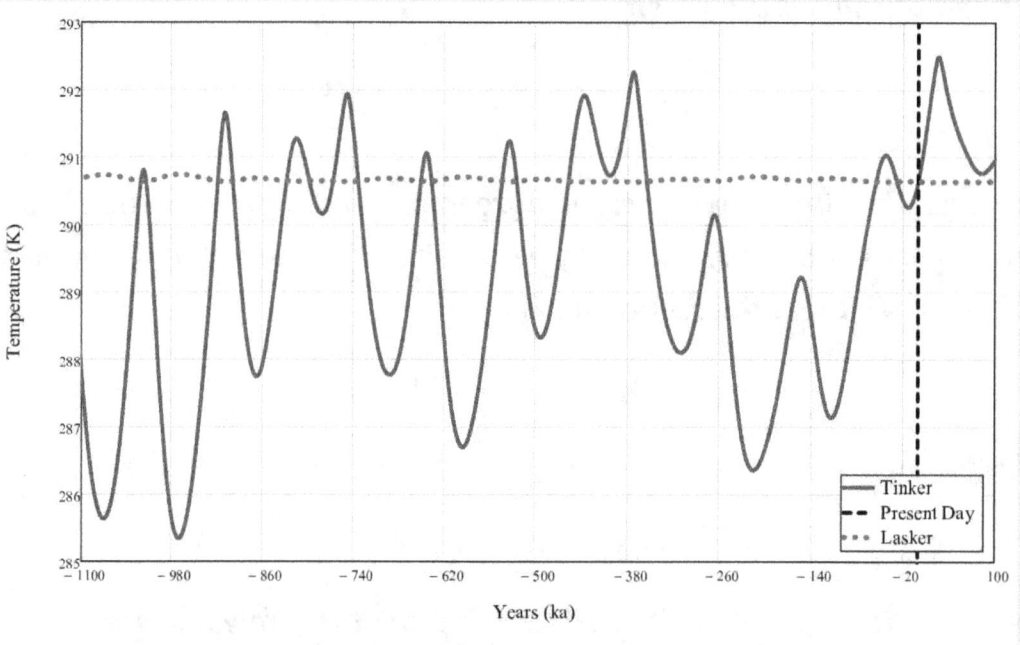

Figure 14 - Temperature using the existing relationship (blue, dotted) versus corrected relationship (red, solid). Vertical dashed line indicates present day.

The temperature span of the revised insolation/eccentricity relationship is vastly greater than that of the relationship presently in use. When this corrected relationship is used, it is shown that the Milankovitch orbit-forcing theory is incontrovertible.

CHAPTER 1 shows conclusively that the Greenhouse Effect theory was born of a mathematical mistake which violates the superposition principle and the fundamental physical law of energy conservation. The surface temperature of the Earth is fully described by properly accounting for solar and geothermal fluxes which means that there is no atmospheric contribution to Earth's surface temperature other than albedo. Simply put, there is no such thing as a Greenhouse Gas.

What, then, of the oft-referenced paleoclimate warm periods under intense study in order to better understand the effect of these mislabeled gasses? Nasa provides a useful statement [15], "When scientists tried to build climate models, they could not get the models to simulate past climate change unless they also added changes in carbon dioxide levels. Though scientists aren't sure why carbon dioxide levels changed, almost all believe that the shift contributed to altering the climate."

Consider the paper of Lasker, et al., "Orbital, precessional, and insolation quantities for the Earth from -20Myr to +10Myr" [16]. The authors generously provide, "...for completeness and clarity...", the techniques used to calculate insolation (solar flux) quantities. These techniques are described as "now classical" (considered settled science) with references to papers of note.

Equation 13 of that paper provides the "now classical" mean annual, top-of-atmosphere, insolation [17] [18] [19] [16] of

$$W_{am} = \frac{S}{4} \cdot (1 - e^2)^{-1/2} . \qquad 2.1$$

Where S is the solar constant and e is the orbit eccentricity.

Despite the consensus behind the "now classical" statement, this relationship is wrong. In this view, as eccentricity increases, so, too, does the insolation. In higher eccentricity orbits, the planet spends more time further from the sun. Therefore, the expected behavior is for insolation to be reduced as eccentricity increases, not the reverse.

The following will derive the correct relationship.

The construction of SECTION C, CHAPTER 1 assumes a constant value for the semi-major axis and the eccentricity of Earth's orbit. It typically takes millennia for these parameters to change appreciably. So, we were justified to make that assumption for the present orbit of the planet. However, to explore mean surface temperatures over long periods, it is necessary to focus on how these parameters change over time.

If we set the eccentricity, e, in Equation 1.24 to zero, we obtain the simplest orbit profile. That is, a circle. In this orbit, the centripetal force exactly matches the gravitational force between a planet and the sun:

$$\frac{GMm}{r_0^2} = \frac{mv^2}{r_0}.$$

2.2

Where G is the gravitational constant, M is the mass of the sun, m is the mass of the planet, v is the planet velocity and r_0 is the circular orbit radius. This circular orbit radius, due only to the planet angular momentum and the sun's gravitational parameter, will be defined as the Keplerian Circular Orbit (KCO). All derived parameters of the KCO will be designated with a subscript of zero, e.g. Ψ_0, etc.

The angular momentum of the planet's motion around the sun is given by

$$L = m \cdot \vec{r} \times \vec{v}$$
$$= m \cdot r_0 \cdot v.$$

2.3

The last step assumes the circular orbit where the cross product is simple multiplication. Applying this to Equation 2.2, we arrive at a description of the circular orbit radius.

$$\frac{mv^2}{r_0} = \frac{L^2}{m \cdot r_0^3} = \frac{GMm}{r_0^2}. \qquad \text{2.4}$$

$$r_0 = \frac{L^2}{GMm^2}$$
$$\equiv \frac{h^2}{GM} \qquad \text{2.5}$$
$$= \frac{h^2}{\mu}.$$

Where h is known as the angular momentum per unit mass and μ is known as the sun's *gravitational parameter*.

The energy of an orbit is derived via the *virial theorem* [9] as being

$$E = -\frac{GMm}{2a}$$
$$= -\frac{\mu \cdot m}{2a}. \qquad \text{2.6}$$

$$\frac{E}{m} \equiv \epsilon$$
$$= -\frac{\mu}{2a}. \qquad \text{2.7}$$

$$\epsilon_0 = -\frac{\mu}{2r_0}. \qquad \text{2.8}$$

Where a is the semi-major axis or, equivalently, in the case of a circular orbit, the circular orbit radius. The circular orbit corresponds to the minimum orbit energy for a given angular momentum. The symbol, ϵ, represents the energy per unit mass of the orbit. ϵ_0 represents specific energy of the circular orbit.

Eccentricity Due to Configurational Change in Orbit Energy

Let's analyze small perturbations of the energy of a circular orbit. We'll start with Equations 2.5 and 2.7 for a circular orbit of radius r_0 and perturb the energy per unit mass by some small value, Δ.

$$
\begin{aligned}
r_0 &= \frac{h^2}{\mu} \\
&= \frac{h^2}{-2\epsilon r_0} \\
\therefore r_0 &= \frac{h}{[-2\epsilon_0]^{1/2}} \cdot
\end{aligned}
\tag{2.9}
$$

Perturbing the energy per unit mass by Δ modifies the KCO radius as

$$
\begin{aligned}
a &= \frac{h}{[2(-\epsilon_0 + \Delta)]^{\frac{1}{2}}} \\
&= \frac{h}{\left[-2\epsilon_0\left[1 - \frac{\Delta}{\epsilon_0}\right]\right]^{\frac{1}{2}}} \\
&\approx \frac{h}{[-2\epsilon_0]^{\frac{1}{2}}}\left[1 + \frac{\Delta}{2\epsilon_0}\right] \\
&= r_0\left[1 + \frac{\Delta}{2\epsilon_0}\right],
\end{aligned}
\tag{2.10}
$$

which I've defined as the Perturbed Keplerian Orbit (PKO) radius, or the semi major axis length, a.

So, in this model, the circular orbit radius is increased by increasing the energy of the system. If the energy of the system is reduced, the circular orbit radius is also.

The important initial condition is that the planet starts its radial oscillation at the original KCO radius of r_0. That is, the KCO radius becomes periapsis after perturbation. If the energy is reduced instead of increased, the original KCO becomes apoapsis. For all accessible orbits, KCO (r_0) is an extremum of the radial oscillation. This must be true to conserve both energy and angular momentum in the perturbed system.

Figure 15 shows a graphical schematic of the process.

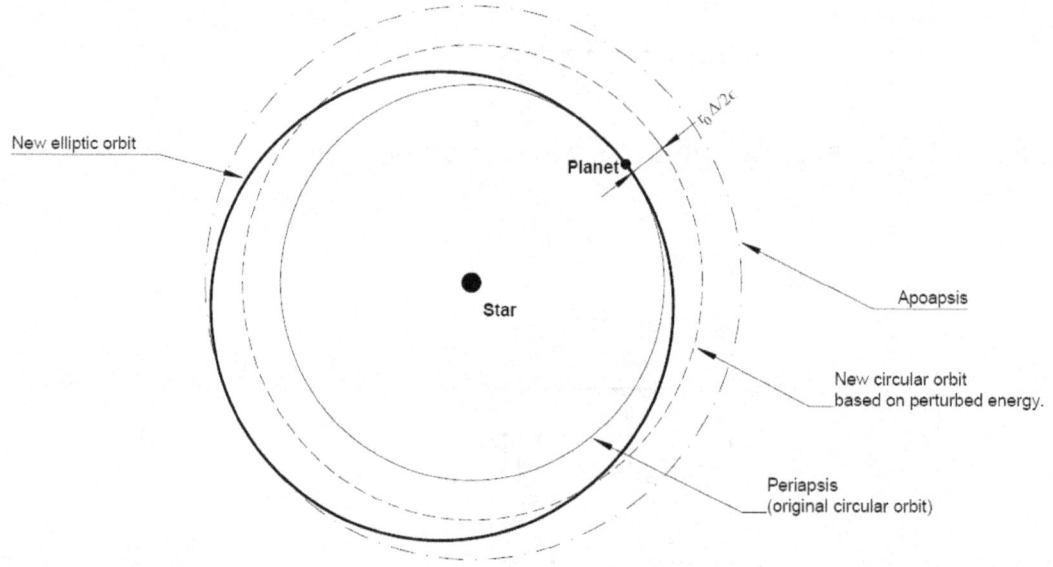

Figure 15 - Perturbed Energy Orbit Change

The eccentricity is readily identified in Equation 2.10 as $\Delta/2\epsilon_0$. However, we must be careful. This is in terms of the KCO and, therefore, is not useful. The actual eccentricity should be in terms of the new semi major axis length, which is identical to the new circular orbit (PKO), radius a. From Figure 15 and Equation 2.10, we can see the relevant eccentricity is given by

$$e = \frac{r_0 \frac{\Delta}{2\epsilon_0}}{a}$$

$$= \frac{r_0 \frac{\Delta}{2\epsilon_0}}{r_0 \left[1 + \frac{\Delta}{2\epsilon_0}\right]}$$

$$= \frac{\frac{\Delta}{2\epsilon_0}}{\left[1 + \frac{\Delta}{2\epsilon_0}\right]}$$

$$= \left[\frac{2\epsilon_0}{\Delta} + 1\right]^{-1}.$$

2.11

As stated previously, Equation 1.24 is the known, *exact* solution to the two-body Kepler problem [9]. Since our perturbation is limited to simply modifying the energy of the orbit, it is reasonable to assume that the new orbit will be accurately described by that same equation. That is,

$$R(\chi) = \frac{a \cdot (1 - e^2)}{1 - e \cdot \cos(\chi)}$$

$$= \frac{a \cdot (1 - e) \cdot (1 + e)}{1 - e \cdot \cos(\chi)}$$

$$= \frac{r_0 \cdot (1 + e)}{1 - e \cdot \cos(\chi)}.$$

2.12

Where the algebraic steps are shown in detail (reference Figure 15) for clarity. In particular, the relationship

$$r_0 = a \cdot (1 - e).$$

2.13

As expected, perihelion and aphelion are as shown in Figure 15,

$$R(0) = \frac{r_0(1+e)}{(1-e)}$$
$$= a \cdot (1+e)$$

2.14

and

$$R(\pi) = r_0.$$

2.15

SECTION B. ORBITAL ENERGY CHANGE DUE TO OTHER PLANETS

The orbital energy perturbation considered in the previous section is quite obviously due to the presence of the other solar system bodies. All of these are largely in nearly circular orbits about the sun at varying distances from it. In truth, they are in elliptical orbits about the center of mass of the solar system, but we lose little by assuming the simpler heliocentric view. So, consider the forces of two planets in different circular orbits on each other.

The gravitational force between the two planets is easily written as

$$\vec{F}_{A-B} = \frac{-Gm_A m_B}{\left|\vec{R}_A - \vec{R}_B\right|^3} \cdot \left[\vec{R}_A - \vec{R}_B\right].$$

2.16

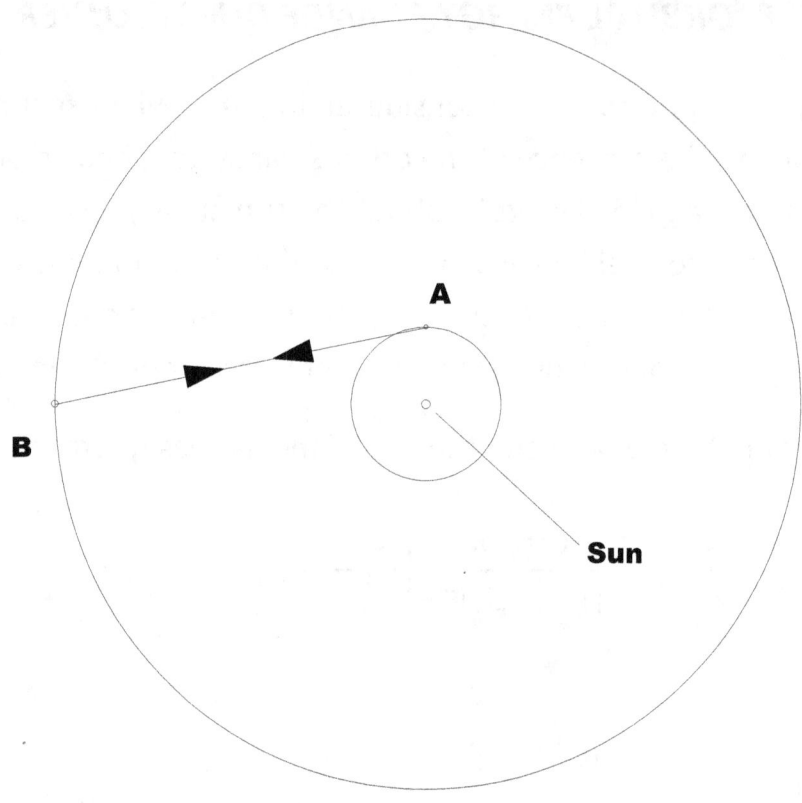

Figure 16 - Planet-A, Planet-B Force

It is instructional to look at this force broken into tangential and radial components. That is, force parallel to the orbit and force perpendicular to the orbit. Focusing on Planet-A,

$$\vec{F}_{A(Rad)} = \frac{-Gm_Am_B}{\left|\vec{R}_A - \vec{R}_B\right|^3} \cdot \left[\vec{R}_A - \vec{R}_B\right] \cdot \vec{R}_A$$

$$= \beta \cdot [R_{A0} - R_{B0} \cdot \cos[(\omega_A - \omega_B) \cdot t]], \qquad \text{2.17}$$

$$\vec{F}_{A(Tang)} = \frac{-Gm_Am_B}{\left|\vec{R}_A - \vec{R}_B\right|^3} \cdot \left[\vec{R}_A - \vec{R}_B\right] \times \vec{R}_A$$

$$= \beta \cdot [-R_{B0} \cdot \sin[(\omega_A - \omega_B) \cdot t]]. \qquad \text{2.18}$$

$$\beta = \frac{Gm_Am_B}{\left[R_{A0}^2 + R_{B0}^2 - 2R_{A0}R_{B0}\cos\left[(\omega_A - \omega_B) \cdot t\right]\right]^{1.5}} \qquad \text{2.19}$$

where R_{*0} is the orbit radius and ω_* is the angular velocity.

Plotting the tangential component of the force reveals,

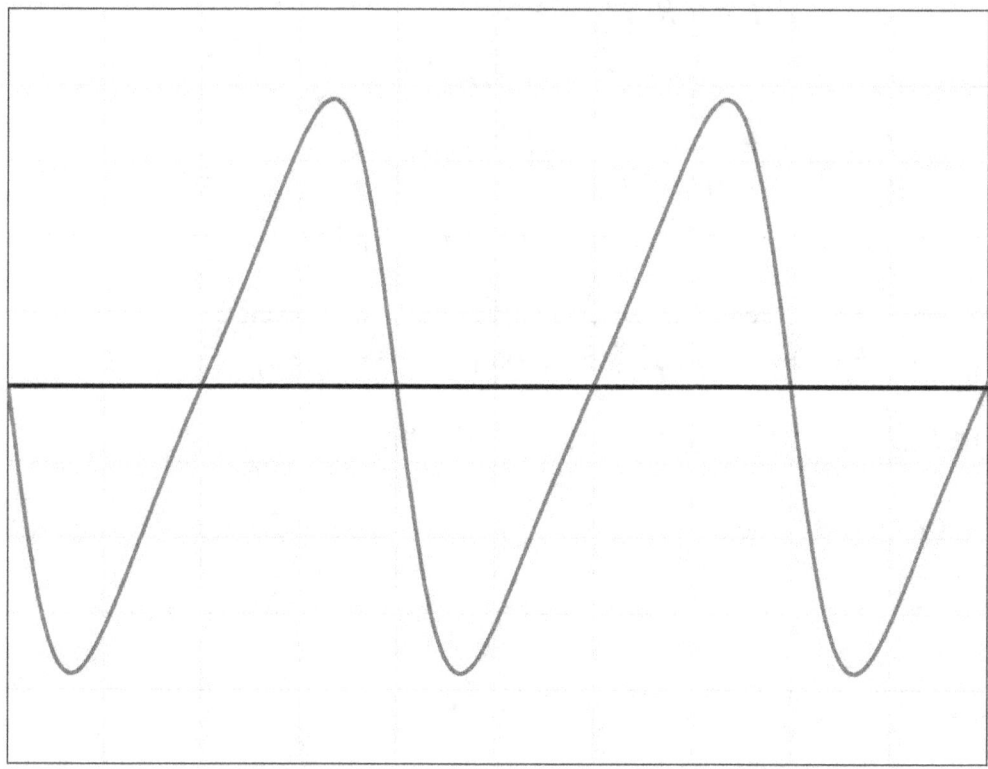

Figure 17 - Tangential Force on Planet-A from Planet-B.
Heavy horizontal line indicates the mean force over time (=0).

The horizontal line is the mean tangential force and has a value of zero. This indicates that the force serves to slightly increase and then decrease the orbital speed. There is no net change in that orbital speed meaning that we can assume an essentially fixed angular momentum.

As found earlier, the new energy-perturbed orbit will have its aphelion or perihelion correspond to the circular, energy-unperturbed orbit. Since the angular momentum is essentially fixed, that radial extremum will have the same angular momentum as it would if the orbit had not been perturbed. It *must* be this way since there is no net tangential force on the planet that would modify the angular momentum.

The radial component of the force plots as

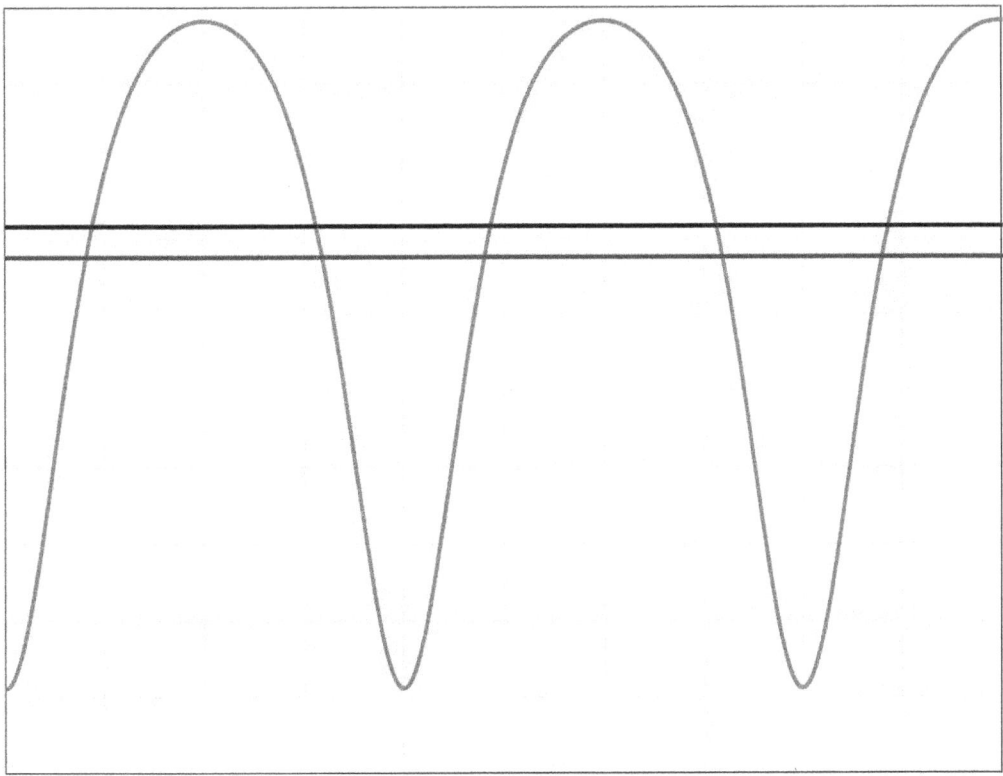

Figure 18 – Radial Force on Planet-A from Planet-B.
The lower heavy horizontal line indicates the mean radial force over time.
Values less than zero (upper horizontal line) indicate force in the
opposite direction of the sun's gravitational attraction.

The lower horizontal line indicates the mean radial force over time. It has a magnitude less than zero indicating that it is in the opposite direction of the gravitational force from the sun. This mean force is the source of the energy change, Δ, in the orbital energy as previously described. This change in energy, as discussed earlier, is responsible for the simultaneous and related change in the semi-major axis and eccentricity values.

As stated previously, Equation 2.1 is inconsistent with its prediction of mean annual insolation when compared with the reality of planets spending more time further from the sun as eccentricity increases. Therefore, a new relationship will be derived from the analysis shown in the preceding few pages, starting with the total power emission from the sun. That emission is represented in Equation 1.46 as $P^\odot$ with the widely accepted, nearly constant value of $3.826 \times 10^{26} W$.

If we imagine a sphere centered on the sun with a radius equal to the radial distance of our prototypical planet from the sun, then the solar emission power per unit area at that orbit point will be given by

$$\Phi(X) = \frac{P^\odot}{4\pi \cdot R^2(X)} \qquad 2.20$$

The dwell-time at a specific value of the angle X, can be deduced from the specific angular momentum, h, which has been shown to be constant both due to the central force aspect of the sun's gravity and due to the zero-mean tangential force from any other planets. That dwell time is given by

$$\frac{\Delta t}{\Delta X} = \frac{R^2(X)}{h}. \qquad 2.21$$

The energy deposited per unit area at the orbit point for that dwell time is

$$\Xi(X) = \frac{P^{\odot}}{4\pi \cdot R^2(X)} \cdot \frac{\Delta t}{\Delta X}$$

$$= \frac{P^{\odot} \cdot R^2(X)}{4\pi h \cdot R^2(X)} \qquad \text{2.22}$$

$$= \frac{P^{\odot}}{4\pi h}$$

The total energy deposited throughout an orbit is obtained by integration,

$$E_{dep} = \int_0^{2\pi} \Xi(X) \cdot dX$$

$$= \frac{P^{\odot}}{4\pi h} \cdot \int_0^{2\pi} dX \qquad \text{2.23}$$

$$= \frac{P^{\odot}}{2h}$$

The mean flux is just this total deposited energy expression divided by the new (perturbed) orbital period. The orbital period is determined by the fixed angular momentum. From Kepler's third law, the change in area, A, in unit time is given by

$$\frac{dA}{dt} = \frac{h}{2}. \qquad \text{2.24}$$

Since the angular momentum is fixed,

$$T = \frac{2A}{h}. \qquad \text{2.25}$$

The area of an ellipse, in terms of eccentricity is

$$A = \pi a^2 \sqrt{(1 - e^2)}$$

$$= \frac{\pi r_0^2 \sqrt{(1 - e^2)}}{(1 - e)^2}$$

$$= \pi r_0^2 \sqrt{\frac{(1 + e)}{(1 - e)^3}} \qquad\qquad \textbf{2.26}$$

$$\equiv A_0 \sqrt{\frac{(1 + e)}{(1 - e)^3}}$$

Where A_0 is the area of the KCO. This provides the period as

$$T = \frac{2A_0}{h} \sqrt{\frac{(1 + e)}{(1 - e)^3}}$$

$$= \frac{2\pi r_0^2}{\sqrt{r_0 \cdot \mu}} \sqrt{\frac{(1 + e)}{(1 - e)^3}} \qquad\qquad \textbf{2.27}$$

$$= 2\pi \sqrt{\left[\frac{r_0}{(1 - e)}\right]^3 \cdot \left[\frac{(1 + e)}{\mu^\grave{}}\right]}$$

$$\equiv T_0 \cdot \sqrt{\frac{(1 + e)}{(1 - e)^3}}$$

Where T_0 is the period for the circular orbit at r_0. Note that the penultimate step in Equation 2.27 is at odds with both Kepler and Newton's relationship between orbit period and semi major axis. This is explored in depth in CHAPTER 4.

Continuing, the mean flux, Q_{sol}, is then given by

$$
\begin{aligned}
Q_{sol}(e) &= \frac{E_{dep}}{T'} \\
&= \frac{P^{\odot}}{2h} \cdot \frac{h}{2A} \\
&= \frac{P^{\odot}}{4A_0} \cdot \sqrt{\frac{(1-e)^3}{(1+e)}} \\
&= \frac{P^{\odot}}{4\pi \cdot r_0^2} \cdot \sqrt{\frac{(1-e)^3}{(1+e)}} \\
&\equiv S_0 \cdot \sqrt{\frac{(1-e)^3}{(1+e)}}.
\end{aligned}
$$

2.28

Collapsing the circular orbit solar constant in the last step is justified since that circular orbit radius is identified as a constant common to all findings of this section.

This flux is intercepted by the disk made by the intersection of the spherical shell about the sun at the orbit radius and the (assumed) spherical planet. This 'slice' through the planet produces a disk with the planet's radius, $\mathcal{R}$, having the area $\pi \cdot \mathcal{R}^2$. The flux so intercepted is distributed over the entire surface area of the planet, $4\pi \cdot \mathcal{R}^2$.

So, the flux incident per unit area on the planet surface (or, more correctly, at top of atmosphere), would be Equation 2.28, multiplied by the planet disk area, then divided by the planet surface area, or,

$$W_{am}(e) = S_0 \cdot \sqrt{\frac{(1-e)^3}{(1+e)} \cdot \frac{\pi \cdot \mathcal{R}^2}{4\pi \cdot \mathcal{R}^2}}$$

$$= \frac{S_0}{4} \cdot \sqrt{\frac{(1-e)^3}{(1+e)}}.$$

2.29

Comparing Equation 2.29 with that used by *Lasker, et al.* (Equation 2.1) shows a vast difference. This reveals the magnitude of the fundamental flaws in the "now classical" technique of determining insolation. This new relation uses the fixed angular momentum of the orbit to define a "solar constant" that is truly constant (has no dependence on a variable semi major axis) since the associated circular orbit is fixed by the angular momentum.

In the referenced paper [16], *Lasker, et al.* did painstaking work in computing Earth's orbit eccentricity as a function of time. Those calculations, while not adhering to the change in semi-major axis discussed herein, did the equivalent of an adiabatic-invariance analysis [9] of eccentricity changes while holding energy constant. Their result is, therefore, reliable to that extent.

Surface temperate can be obtained by applying albedo effects as described in Equation 1.47 to top-of-atmosphere insolation and then combining the result with geothermal flux as prescribed by Equation 1.8. Thus,

$$T_L(e) = 64.8 \left[\left[\frac{0.71 \cdot S}{4} \cdot (1 - e^2)^{-\frac{1}{2}} \right]^{0.25} + 0.087^{0.25} \right]. \qquad \text{2.30}$$

$$T_T(e) = 64.8 \left[\left[\frac{0.71 \cdot S_0}{4} \cdot \sqrt{\frac{(1-e)^3}{(1+e)}} \right]^{0.25} + 0.087^{0.25} \right] \qquad \text{2.31}$$

Where the subscripts L and T on the temperature T indicate "*Laskar, et al.*" and "*Tinker*", respectively.

As can be easily seen, the temperature excursion using Equation 2.31 far exceeds that of *Lasker, et al.* Equation 2.30. It is important to note the rapidly rising leading edge of the temperature peak we are currently experiencing. The peak of that excursion should occur (according to *Lasker, et al.* eccentricity calculations) in about *25,000* years at about *292.5K* (*19.35C*). That temperature will be the hottest global mean temperature in over a million years.

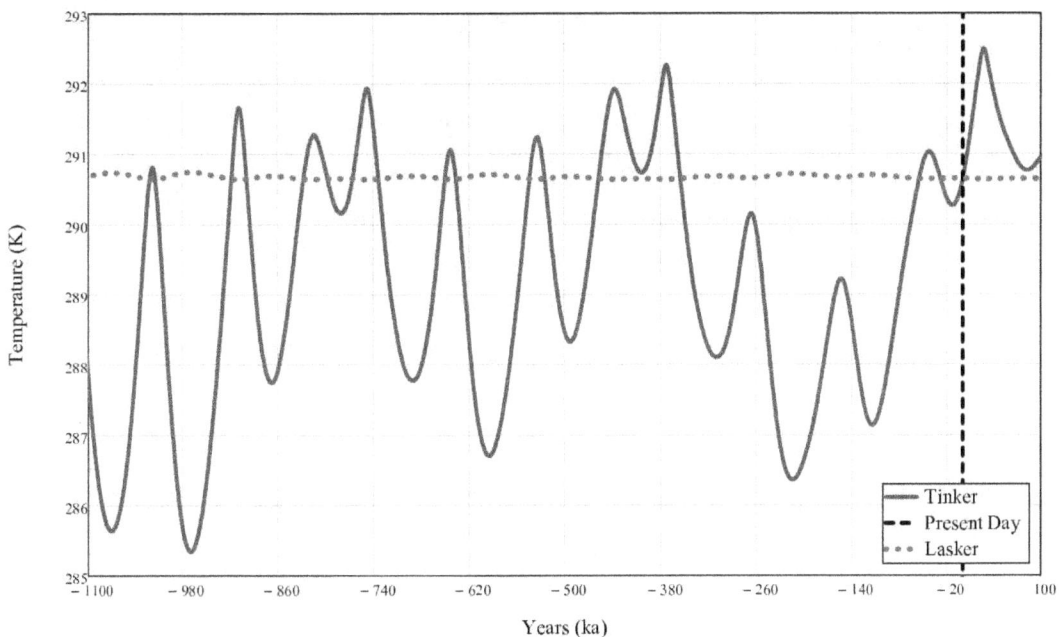

Figure 19 - Temperature reconstruction using this work's Equation 2.31 (red, solid) and that of *Lasker, et al.* Equation 2.30 (blue, dotted). The vertical dashed line indicates the present time on the graph.

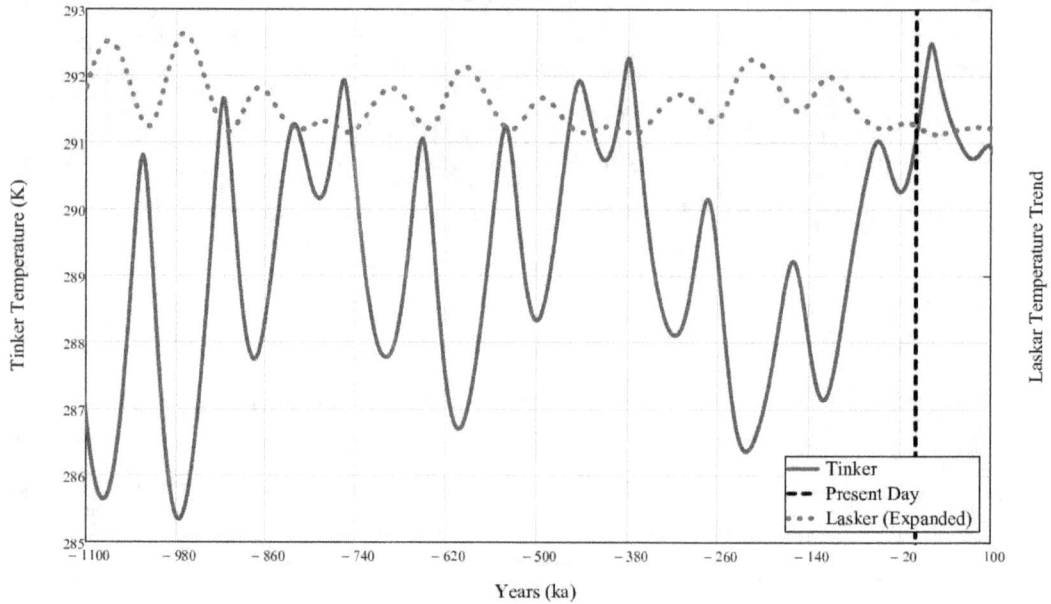

Figure 20 – Same as Figure 19 with *Laskar, et al.* temperature reconstruction expanded. This shows that the *Laskar, et al.* reconstruction is inverted with respect to the (expected) temperature variations of *Tinker*.

Chapter 3 - Executive Summary

This chapter uncovers the failure of researchers to properly analyze the kinetic fractionation process on which historical temperatures are deduced from ice core data. To this, researchers obtained minor water-molecule-isotope abundance data from ice cores near the poles. This provided isotope abundance as a function of time derived from the depth of each sample. Direct temperature data was also obtained for a representative number of bore-hole depths.

The error made at this point was to try to construct a linear fit of temperature to the isotope concentration measurements. This is at odds with the kinetic fractionation mechanism that the data probes. Researchers erroneously assumed that the linear fit to temperature was a reasonably reliable proxy for paleoclimate temperatures when, instead, the fractionation process is dominated by the time change in temperature, dT/dt. This is shown conclusively by plotting the flawed temperature/isotope proxy assumption against the temperature reconstruction of the last chapter and then doing the same with the derivative of that temperature reconstruction with respect to time. These are shown in the following two images. The isotope data is a remarkable match to dT/dt reconstruction while having little relationship to the temperature reconstruction.

This indicates that researchers using the linear fit of temperature to isotope data have been employing wildly inaccurate temperature data in their hypotheses, theories, and analyses. Further, since temperature dependence of the solubility of gases in water follows the same mathematical derivation as kinetic fractionation, trace gas concentrations

61

in ice cores simply act as another proxy for dT/dt. They do not drive temperature, as we learned in CHAPTER 1.

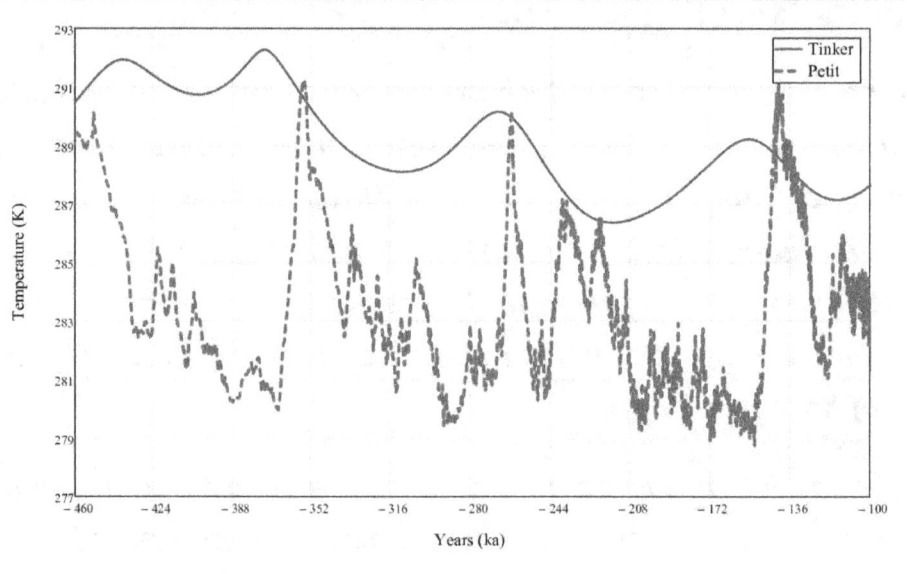

Figure 21 - Comparison between computed temperature (red, solid) and isotope ratio abundance measurements (blue, dotted).

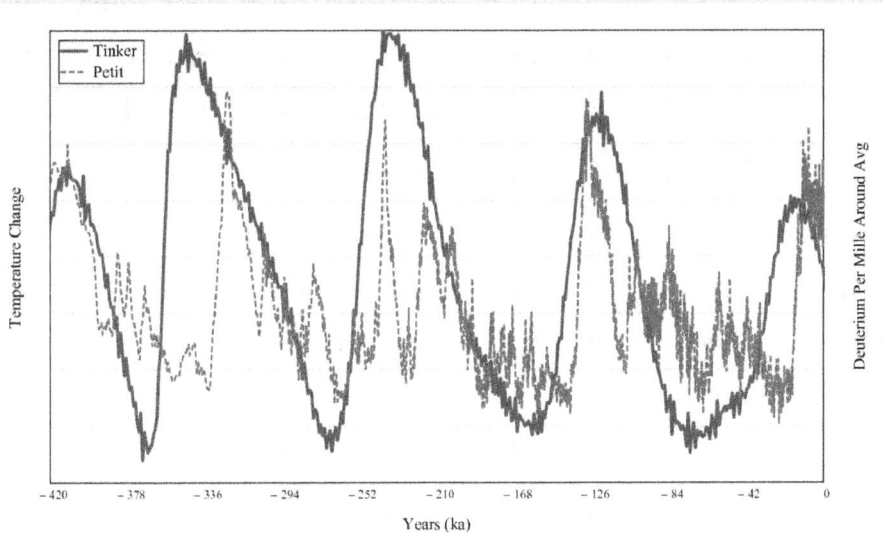

Figure 22 – Time derivative of CHAPTER 2 temperature reconstruction (red, solid) together with measured isotope ratio abundance (blue, dotted).

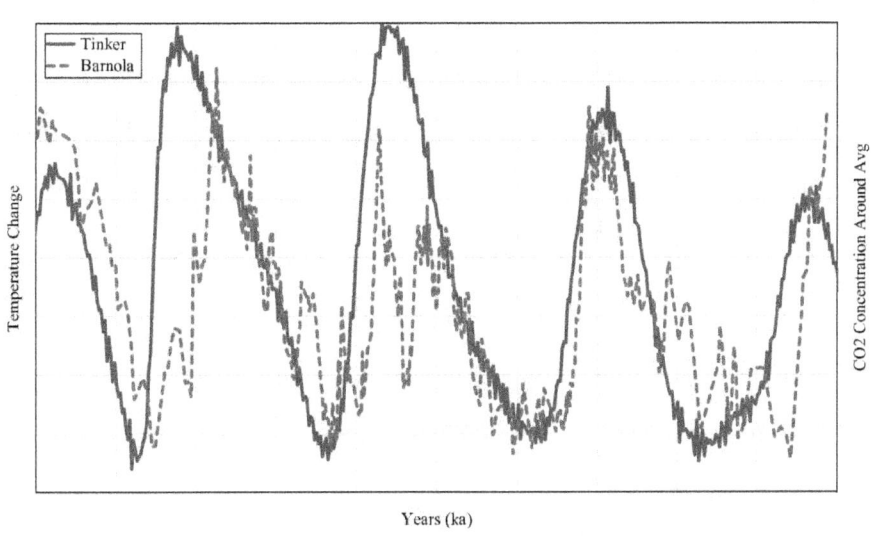

Figure 23 – Time Derivative of CHAPTER 2 temperature reconstruction (red, solid) together with the CO_2 concentration from *Barnola, et al.* [20] (blue, dotted).

By retrieving ice from great depths, scientists have been able to obtain a record of atmospheric constituencies from the distant past. By various methods, layers of these ice cores can be associated with the approximate date the layer was deposited thereby giving a useful record of atmospheric constituency as a function of time.

One of the interesting molecules trapped in these ice cores is that of water made with an isotope of hydrogen having an additional neutron. The most common isotope of hydrogen has one proton and zero neutrons, written ^1H. The heavier version is written ^2H.

By measuring the ratio of ^2H to ^1H, it is possible to deduce information about the temperature at the time associated with the depth at which the ice was taken. This is because the act of evaporating water from warmer regions, atmospheric transport to colder regions, followed by condensing and precipitation can change the $^2H/^1H$ abundance from what it is/was at the source latitudes.

Scientists also measured ice temperature at various depths along the ice-core shaft. Using the isotope ratio and temperature data, they determined a linear calibration for converting isotope ratio abundance into temperature. A typical calibration is given by

$$T = T_0 + 0.111 * \delta^2 H \qquad\qquad 3.1$$

where

$$\delta^2 H = \frac{\left(\frac{^2H}{^1H}\right)_f - \left(\frac{^2H}{^1H}\right)_i}{\left(\frac{^2H}{^1H}\right)_i} \qquad\qquad 3.2$$

In this, the *i* and *f* subscripts correspond to *initial* ratio taken from "standard mean ocean sea water" and *final* ratio taken from the ice core.

Using ice core data from *Petit, J.R., et al.* [21], and the temperature reconstruction from CHAPTER 2, we get this comparison plot:

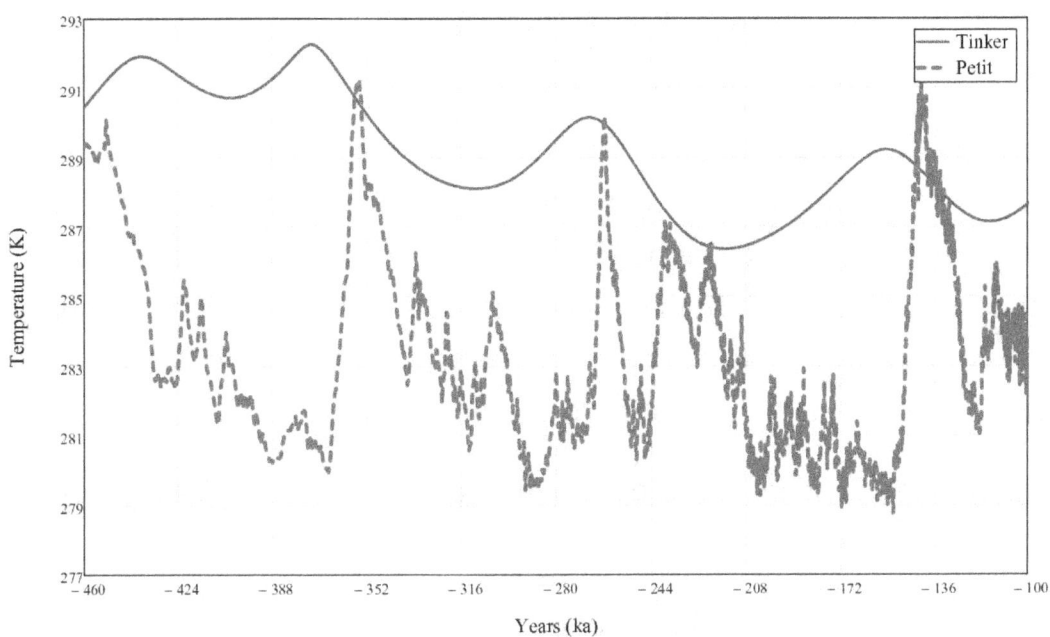

Figure 24 - Comparison between computed temperature (red, solid) and isotope ratio abundance measurements (blue, dotted).

In this, *288K* is the T_0 value used in Equation 3.1. It is easy to see that the peaks of the data fall close to the computed temperatures, but the actual structure of the temperature record is substantially different from the computed values. Let's resolve this by analyzing the process involved.

As mentioned above, evaporation and subsequent condensation of sea water form a $\delta^{18}O$ isotope fractionation process that is detectable in ice core samples. This mechanism of fractionation has a continuum of rates depending on the equilibrium state of the system. Without focusing on the

65

details of the mechanism it is still germane to assume a Taylor-series expansion of the fractionation rate, f, as

$$\left(\frac{^2H}{^1H}\right)_f = \left(\frac{^2H}{^1H}\right)_i f[T(t)]$$

$$\approx \left(\frac{^2H}{^1H}\right)_i \left[f(T_0) + f'(T_0)\frac{\partial T}{\partial t}\right].$$

3.3

where the f subscript indicates the final isotope ratio and i the initial source isotope ratio as described above. The process temperature, T, is assumed to be the mean surface temperature of the planet. At T_0 the system is assumed to be in equilibrium with respect to the fractionation mechanism. That is, the isotope abundance ratio in the ice is the same as the isotope abundance ratio in the mid-latitude source waters.

As a reminder, ice core isotope ratio measurements are typically given as

$$\delta^2 H = \frac{\left(\frac{^2H}{^1H}\right)_f - \left(\frac{^2H}{^1H}\right)_i}{\left(\frac{^2H}{^1H}\right)_i}$$

$$= f(T) - 1$$

3.4

$$\approx (f(T_0) - 1) + f'(T_0)\frac{\partial T}{\partial t}$$

$$\equiv \varsigma + v\frac{\partial T}{\partial t}.$$

where I have detailed the relationship for small changes in temperature about a mean. The last step assumes $f(T)$ to be weakly dependent on T. One can view ς and v as being related to the equilibrium and kinetic fractionation rates, respectively. Differentiating the calculated results shown in Figure 19 and plotting together with Vostok $\delta^2 H$ ice-core data used above, we find

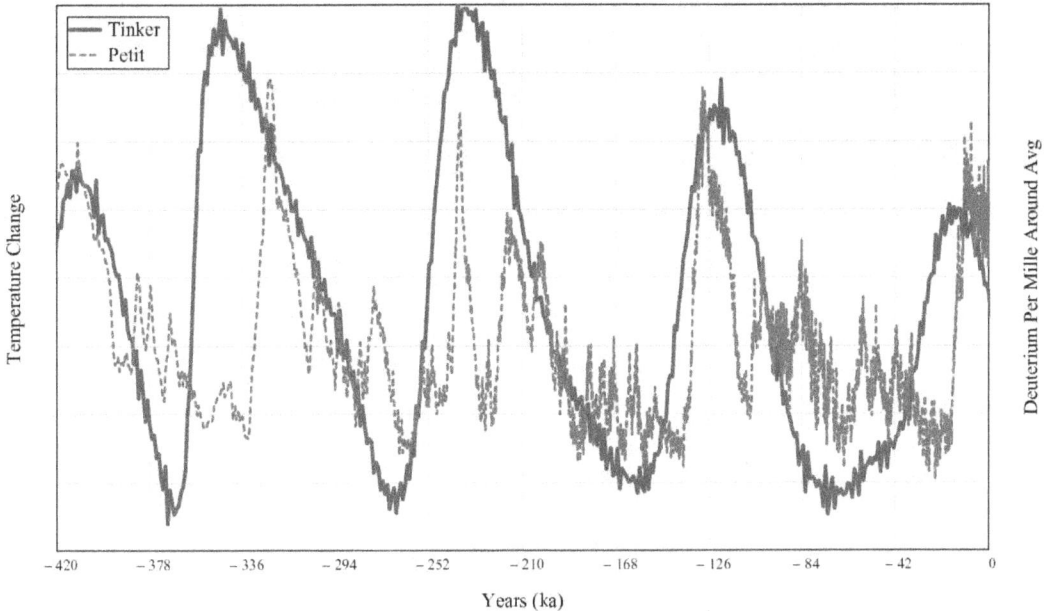

Figure 25 – Time rate of change of the temperature reconstruction
from CHAPTER 2 (red, solid) together with the isotope
ratio abundance from *Petit, et al.* (blue, dotted).

It should be evident from Figure 24 and Figure 25 that the ice core isotope ratio abundance is decidedly not a proxy for temperature, as researchers have reported. It is, instead, a proxy for the change in temperature in time, $\Delta T/\Delta t$. It should also be clear that the temperature reconstruction of CHAPTER 2 is remarkably accurate in reproducing the $\delta^2 H$ ice core data once the analysis is performed correctly. This is testament to the validity of the model and to the accuracy of the *Lasker, et al.* calculations.

Figure 25, clearly shows that the time derivative of the temperature computed using Equation 2.31 provides a remarkable match to the character, amplitude, and temporal structure of the data. Therefore, the temperature profile found using Equation 2.31 is assumed to be an accurate representation of Earth's global mean temperature as a function of time.

Solubility of gases in sea water as a function of temperature is also a well-researched subject of physical chemistry. As temperature increases, the solubility decreases. We can construct a mathematical relationship similar to Equation 3.4 for changes in the abundance ratio of a gas in the atmosphere to its abundance in the ocean:

$$
\begin{aligned}
\delta^2 \varrho &= \frac{\varrho_a - \varrho_o}{\varrho_o} \\
&= g(T) - 1 \\
&\approx (g(T_0) - 1) + g'(T_0)\frac{\partial T}{\partial t} \\
&\equiv \kappa + \lambda\frac{\partial T}{\partial t}.
\end{aligned}
\qquad \text{3.5}
$$

where $g(T)$ is proportional to the solubility of the gas, ϱ, in sea water and ϱ_a, ϱ_o are atmospheric and ocean abundances of that gas respectively. Again, at temperature T_0 the system is assumed to be in equilibrium – the quantity of gas kinetically exiting the ocean equals that entering from the atmosphere.

It is clear that derivation of Equation 3.5 is the same as that of Equation 3.4 meaning that empirical data should reflect the same behavior. A case in point is the abundance of CO_2 in ice core data. Again, differentiating the calculated results shown in Figure 19 and plotting together with changes in Vostok CO_2 ice-core data about the mean, we find what is seen in Figure 26.

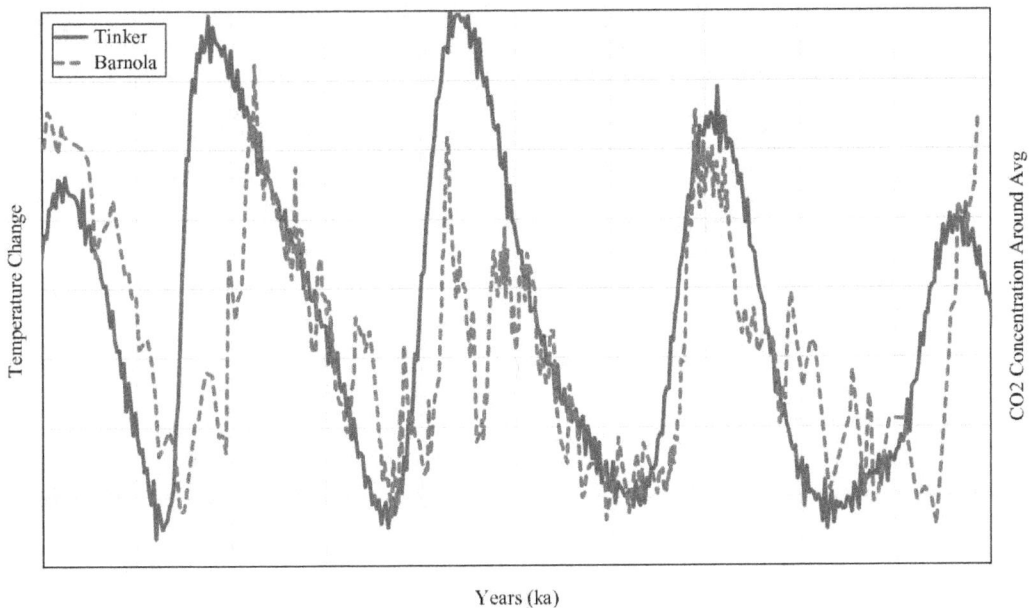

Figure 26 - Derivative of the temperature reconstruction from CHAPTER 2 (red, solid) together with the CO_2 concentration from *Barnola, et al.* [20] (blue, dotted).

Again, since the mathematics are the same, changes in atmospheric abundance of all water-soluble trace gases *in the absence of other sources* will serve as a proxy for $\Delta T / \Delta t$. It should be no surprise that they all reveal a similar form and correspondence to $\Delta T / \Delta t$. They should all have different values for the parameters κ and λ in Equation3.5, but the character should be similar.

Another proxy of note are *foraminifera (forams).* These single-cell organisms live in the oceans, consuming nearby oxygen that is deposited in their shells. As discussed above, the isotopic abundance of ocean-dissolved gasses varies as a function of temperature according to Equation 3.5. This means that measuring the depth and isotopic ratios of oxygen in foram deposits can provide data that is an analog to the Vostok ice core data. This

69

is detailed by *Mortyn, P. G., et al.* [22]. As an analog for ice core data, forams are also a proxy for $\Delta T / \Delta t$ and not a temperature proxy.

In today's academia there is a fervent belief that Figure 27 is the temperature profile over the last *420,000* years [21]. The assumed Last Glacial Maximum is noted by the arrow.

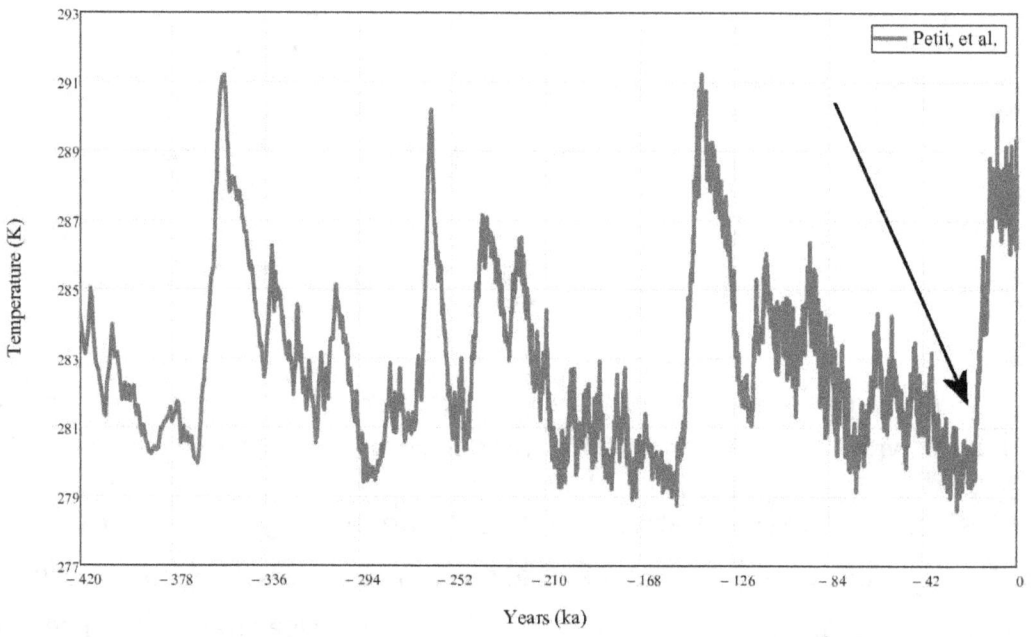

Figure 27 - The present-day assumption of Earth's mean surface temperature over the last 420,000 years. The arrow points at what is considered the Last Glacial Maximum, *20,000* years ago.

However, through the derivation of this chapter, the profile of Figure 27 has been shown to be related to the time rate of change of temperature. The valleys of Figure 27 come about when the fractionation state reaches temporary equilibrium. That is, when the sea water abundance and that in the atmosphere are in kinetic equilibrium. They are not temperature spikes.

Figure 28 properly represents the long-term temperature profile. The "Last Glacial Maximum" assumption is marked there also.

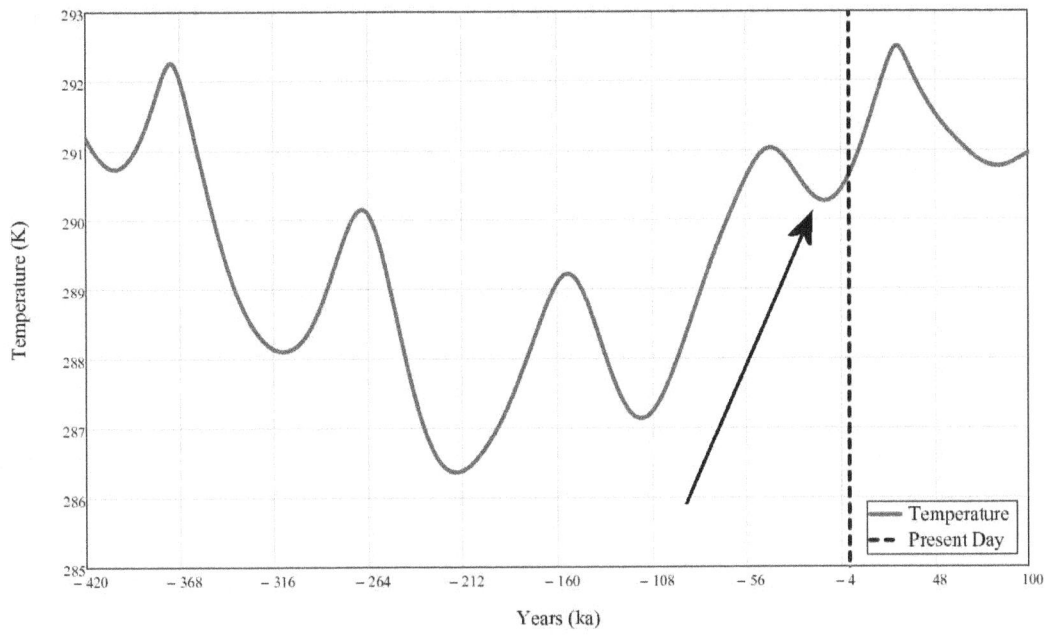

Figure 28 - Last Glacial Maximum...

The profile is expanded in Figure 29 around the last glaciation maximum estimate of ~*-20ka* in Figure 29. The dashed line indicates *-14ka;* zero is the present. The plot shows a global mean temperature about *0.5C* less than today. This is far less than a recent estimate of *6C* [23] where the models used include those found to be invalid herein.

71

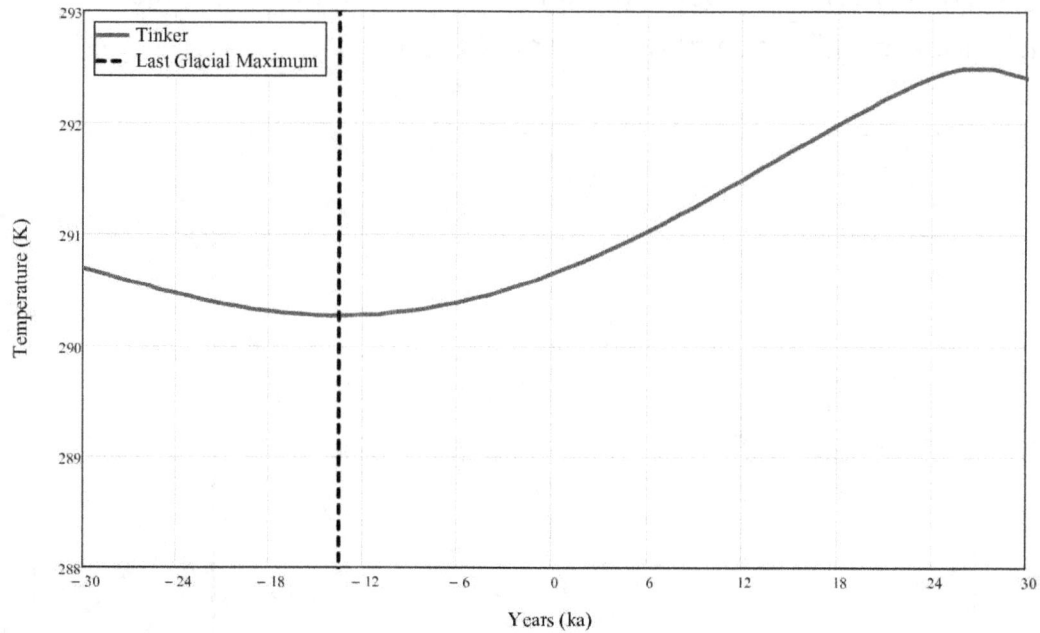

Figure 29 - Temperature profile from -30 to +30 ka. Dashed line shows -14ka.

Chapter 4 - Executive Summary

This explores a proposed change to Kepler's Third Law. The derivation of Equation 2.27 revealed that an eccentricity factor in that law is necessary in order to accurately compare two orbits. It is found that the application of this change to known orbit parameters results in little change for all planets except Mercury and Pluto.

The fact that most planets' orbit parameters are largely unchanged by the proposed modification to the third law indicates that Kepler's observations were not in error. However, Newton showed that modification of the Third Law to include planet masses could make it more accurate and more useful. This proposal to include orbit energy is made in the same spirit.

As illustration of the benefits making this change would realize, it is shown that Mercury's anomalous perihelion precession observed in 1859 by Urbain Le Verriere is fully resolved by the underlying physics that prompt the proposal. This is the only classical physics model that resolves this precession issue. Prior to this, General Relativity had to be employed to find a solution.

Both the solution provided herein, and that of General Relativity are paths to the same destination. They both do the same thing, but they do it from two different perspectives. While the result does not call into question the General Theory of Relativity, it does reveal that General Relativity is not necessary to achieve a solution. As such, the perihelion issue should not be used as verification of General Relativity's applicability.

Astronomers should note the extra factor of $(1 + e)$ in Equation 2.27. Kepler's third law states that the square of an orbit's period is proportional to the third power of the orbit's semi major axis. This is borne out in the penultimate step of the derivation in Equation 2.27, but the proportionality constant differs from that widely used in astronomy. That is,

$$\frac{T^2}{a^3} = \frac{4\pi^2 \cdot (1 + e)}{\mu}.$$

4.1

The conventional statement of this constant is

$$\frac{T^2}{a^3} = \frac{4\pi^2}{\mu}.$$

4.2

That is, the form currently in use does not include the (1+e) factor derived here.

Newton modified Kepler's relationship by noting that two masses orbit about their center of mass and not one about the other.

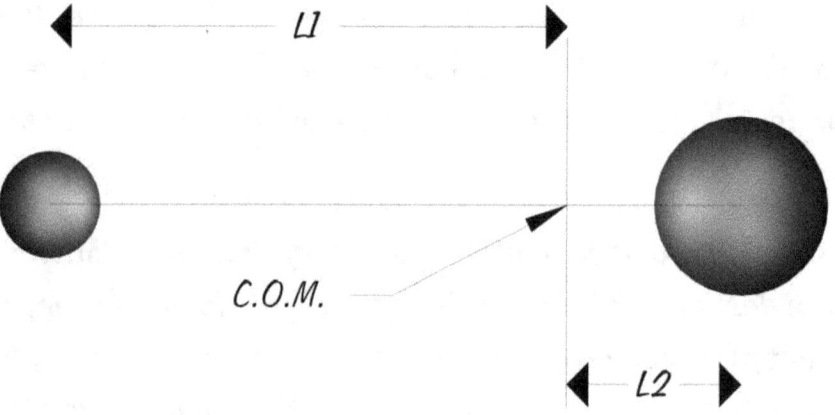

Figure 30 - Newton modification of Kepler.

This results in

$$\frac{T^2}{(L1 + L2)^3} = \frac{T^2}{a^3} = constant$$

$$\left[= \frac{4\pi^2}{G(M + m)} \right].$$

4.3

Where the constant is brackets reflects the commonly assumed Equation 4.2. In this, M is the mass of the large body (the sun) and m is the mass of the small body (planet). However, in an orbit with radial oscillation (eccentricity), both masses oscillate with respect to the center of mass. Therefore, Equation 4.1 still governs and the correct relationship is

$$\frac{T^2}{a^3} = \frac{4\pi^2 \cdot (1 + e)}{G(M + m)}.$$

4.4

This is critically important in that the relationships in Equations 4.2 and 4.3 have been in pervasive use for centuries to determine orbital periods, semi major axis lengths, and planet masses both within our solar system and for exoplanets. This is done using a known orbit to determine the value of the constant in 4.2 or 4.3, then applying it to a different orbit to determine an unknown.

Equations 4.1 and 4.4 are consistent with Kepler's observation and Newton's modification of it for small eccentricities, but the details of the constant value are different than what is widely assumed. Because that constant is shown to include orbit eccentricity, the values of all orbital parameters that have been determined using the constant of Equation 4.2 or 4.3 are suspect.

At present, most planet eccentricities are small and approximately the same as Earth's. For comparison, Table 1 shows the mass, semi-major axis length, and eccentricity values gleaned from the J2000.0 epoch [24].

Planet	Orbit Period (days)	Mass (10^{24} kg)	Semi Major Axis (au)	Eccentricity
Mercury	87.979	0.3302	0.3871	0.2056
Venus	224.70	4.8685	0.7233	0.0068
Earth	365.25	5.9795	1.000*	0.0167
Mars	686.97	0.7349	1.5237	0.0298
Jupiter	4335.47	1898.13	5.2051	0.0498
Saturn	10831.47	568.34	9.5815	0.0556
Uranus	30800.38	86.813	19.230	0.0444
Neptune	60307.62	102.413	30.097	0.0112
Pluto	90679.53	0.0317	39.501	0.2479

Table 1 - NASA JPL HORIZONS J2000 [24] Orbit Parameters (* definition)

Using μ of Equation 1.26, we can calculate the semimajor axis using Equations 4.1 and 4.4 to get

Planet	Orbit Period (days)	Semi Major Axis (au) Eqn. 4.2	Semi Major Axis (au) Eqn. 4.1	Eccentricity
Mercury	87.979	0.3871	0.3637	0.2056
Venus	224.70	0.7233	0.7218	0.0068
Earth	365.25	1.000*	1.000*	0.0167
Mars	686.97	1.5237	1.5090	0.0298
Jupiter	4335.47	5.2051	5.1217	0.0498
Saturn	10831.47	9.5815	9.4101	0.0556
Uranus	30800.38	19.230	18.955	0.0444
Neptune	60307.62	30.097	29.988	0.0112
Pluto	90679.53	39.501	36.693	0.2479

Table 2 - Orbital parameters computed using Equations 4.1 and 4.4. (* definition)

With the exception of Mercury and Pluto, all semi major axis calculations using Equations 4.1 and 4.4 are nearly identical. If we examine the orbit of

Mercury, we find both a verification of Equation 4.1's validity and another question.

In a wonderful, but unpublished review entitled *Mercury's Perihelion* [25], *Chris Pollack* describes the discovery by *Urbain Jean Joseph Le Verriere* of a discrepancy in the precession of the perihelion of Mercury. His observations revealed a precession of *565 arcseconds per century* versus an accurate, but difficult calculation that yields only *527 arcseconds per century*. In his review, *Pollack* reproduces the calculation technique showing an expression for the angle between perihelion and aphelion (paraphrased here),

$$\Psi = \pi \cdot [1 + a \cdot (\beta + \gamma)],$$
4.5

where β and γ are numerically computed values with terms in powers of the semi major axis, a, exceeding first order. Ignoring corrections to these higher order terms, and simply correcting the linear term by substituting $a \longrightarrow (1 + e)^{\frac{1}{3}} \cdot a$, we have a corrected version

$$\Psi_c = \pi \cdot \left[1 + (1 + e)^{\frac{1}{3}} \cdot a \cdot (\beta + \gamma)\right].$$
4.6

Precession, using this revised result, is computed as

$$precession = \frac{2\pi \cdot (1 + e)^{\frac{1}{3}} \cdot a \cdot (\beta + \gamma)}{T}.$$
4.7

T is the orbit period of Mercury. So, using *Le Verriere's* result of *527 arcseconds per century*, we correct with

$$corrected = (1 + e)^{\frac{1}{3}} \cdot 527$$
$$= (1.2056)^{\frac{1}{3}} \cdot 527 \qquad\qquad 4.8$$
$$= 1.064 \cdot 527$$
$$= 566 \; arcseconds \; per \; century.$$

So, the perihelion precession issue found by *Le Verriere* is entirely attributable to the use of the conventional relationship in Equation 4.2. It is resolved by using Equation 4.1 derived here.

Kepler's observations concerning planetary orbits were snapshots of multiple, weakly-coupled systems. What he observed and the solution thereof (Equation 1.24) was exactly how two bodies would orbit their center of mass given their masses and two fixed parameters, angular momentum and energy. The formalism of Hamilton's equations provided the framework for exactly solving that problem since the energy of that system was fixed.

What the development shown in Equation 2.10 does, is to identify a baseline, or reference state in which there is no eccentricity and, consequently, no detectable perihelion precession, i.e., the *Keplerian Circular Orbit*. Further, it derives how that reference system, how the Hamiltonian description of it, will behave if the energy is perturbed. Specifically, it shows that perturbing that Hamiltonian, changing the energy of the system slightly, modifies one single aspect of the system dynamics. It only changes the eccentricity of the orbit. Since I have chosen the circular orbit consistent with the fixed angular momentum, the reference state (*KCO*) is also the minimum energy state of the system. The KCO and the eccentricity allows recovery of all other orbit parameters.

What Equations 4.1 and 4.4 show is how the ratio T^2/a^3 is changed when the Hamiltonian is perturbed from the circular orbit reference state. As a result, this chapter provides a method to analyze the evolution of the system, the orbit, provided one knows the time dependence of the Hamiltonian itself, i.e., $\partial H/\partial t$.

The factor of *(1+e)* in Equations 4.1 and 4.4 is a consequence of defining the *KCO* as the reference state. Without a reference state that has physically identifiable correspondence in all related systems, it is impossible to make an accurate comparison. As such, the findings of this chapter are distinct from Kepler's observations as his were among systems not all of which in an identifiable and corresponding reference state.

That said, Kepler's observations were dominated by orbit reference states nearly equal to the *KCO* reference states. Therefore, the ratios of T^2/a^3 were mostly of the same value. Mercury and Pluto's orbits, however, are in energy states significantly different from their *KCO* reference states and, therefore, their ratios of T^2/a^3 significantly differ from what they would be if their energy states were comparable to the other planets' orbits.

The ratio of

$$\frac{[T^2/a^3]_i}{[T^2/a^3]_j} = \frac{(1+m_j/M)\cdot(1+e_i)}{(1+m_i/M)\cdot(1+e_j)} \qquad 4.9$$

Where *i* and *j* identify two different orbits, simply indicates the differences in the masses and the energy states. The same way that Newton modified Kepler's third law to include masses, this chapter modifies Newton's correction to include the orbit energy (referenced to *KCO*) represented by the eccentricity.

The "other question" alluded to earlier is that the first success of the theory of General Relativity was to solve the Mercury perihelion precession issue. So, the question is, does my result nullify the General Relativity result or vice-versa?

The answer is, "Neither". What my result shows is that General Relativity is not necessary in order to resolve the anomalous precession rate of Mercury's orbit. What it shows, specifically, is that the precession rate is not anomalous if one applies the total energy change in an orbit referenced to an energy state in which there is no precession. That total energy change is what is responsible for eccentricity and precession. By choosing a reference state (KCO) in which there is no precession and referencing "energy change" to that, the precession rate is readily determined.

Apparently, applying General Relativity to the problem is akin to finding the total energy change in the orbit referenced to a state in which there is no eccentricity. In this case, the state of no eccentricity would be that of bare, empty space. In the same way that the KCO reference revealed energy not accounted for when using a different reference state, General Relativity was equally successful.

So, there is no abnormality in both General Relativity and the orbit model herein finding a solution to the Mercury perihelion anomaly. They both find the same solution in different ways. What is of note, is that the solution of the Mercury perihelion anomaly is not a success that necessarily supports the Theory of General Relativity. General Relativity is not necessary to find the solution.

Chapter 5 - Executive Summary

This chapter provides data on "energizing events" in science that lead to a marked increase in the rate of published papers in the field. Such an event occurred in material science when Krätschmer, et al. published their paper on manufacturing bulk quantities of fullerenes (Carbon-60).

Another such event occurred in global warming/climate change that, unlike the fullerene case, was due to government funding as opposed to a scientific breakthrough.

Further, the chapter explores the possible issues with politically guided scientific funding. It is suggested that such biased funding policies have a deleterious effect on scientific inquiry. It concludes that this was the case in climate change research.

I had the privilege to work with Professor Donald Huffman at the University of Arizona around the time he and Wolfgang Krätschmer made public their discovery of how to create bulk quantities of carbon-60 (buckminsterfullerene) [26]. In that, I was tasked with maintaining a database of papers on the general subject of fullerenes. It was striking how that discovery launched intense focus on the study of all aspects pertaining to this new form of carbon.

If we examine the aggregate number of papers published on fullerenes as a function of time, we can easily see the almost immediate increase in the publication rate that the Krätschmer /Huffman paper spawned (Figure 31). This is not a typical picture of scientific progress.

Most fields of study exhibit the behavior in the earlier part of that plot. There is a slow, methodical increase in knowledge about the area of interest. As you can see by the later portions of the curve, the progress is still steady, but more rapid. It is the abrupt change in the rate that is the sign of an energizing event.

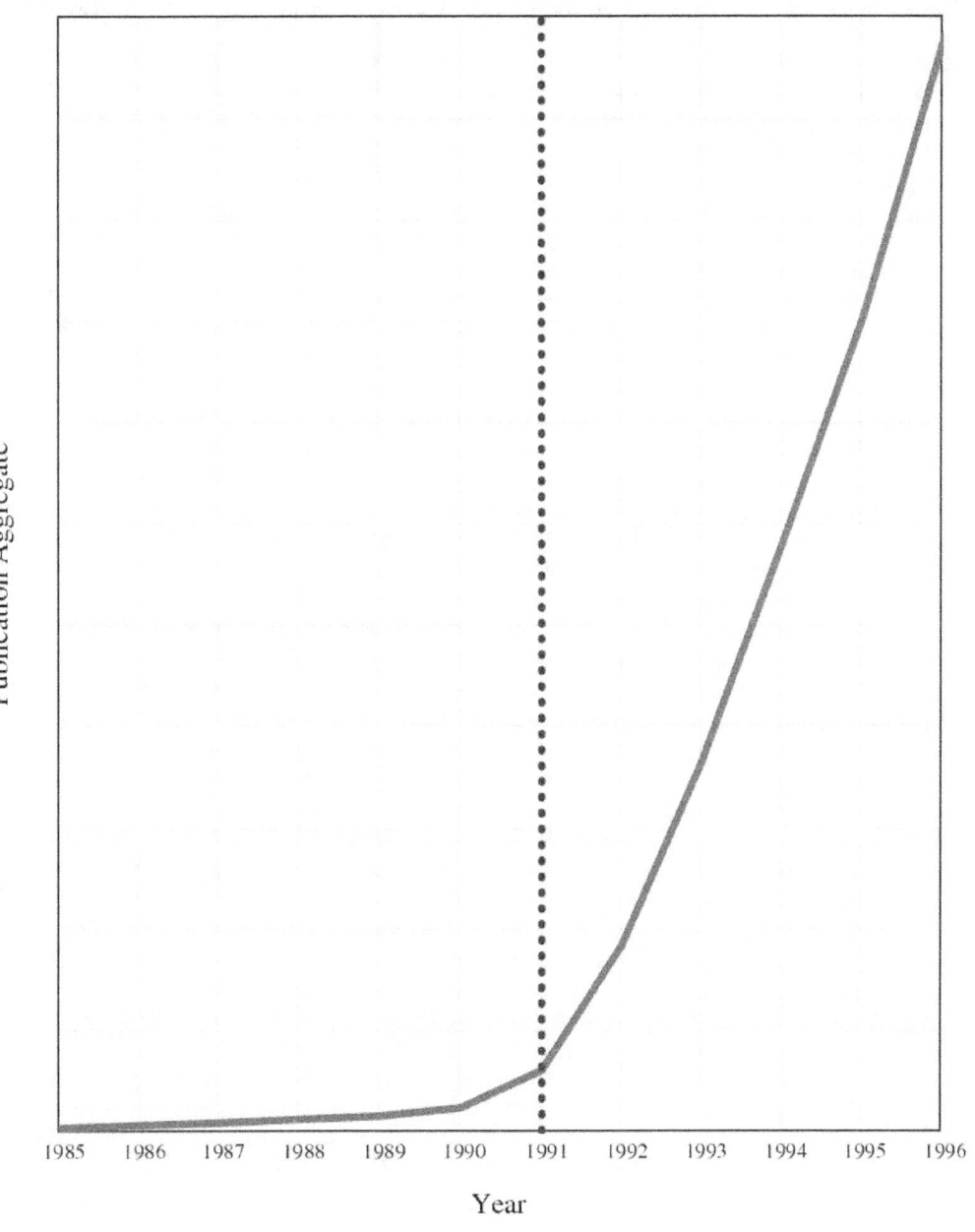

Figure 31 - Krätschmer/Huffman influence on fullerene investigation. The dashed
vertical line indicates the publication date of their paper.

If we examine the same publishing behavior in the field of Global Warming/Climate Change, the same sort of energizing event is also evident.

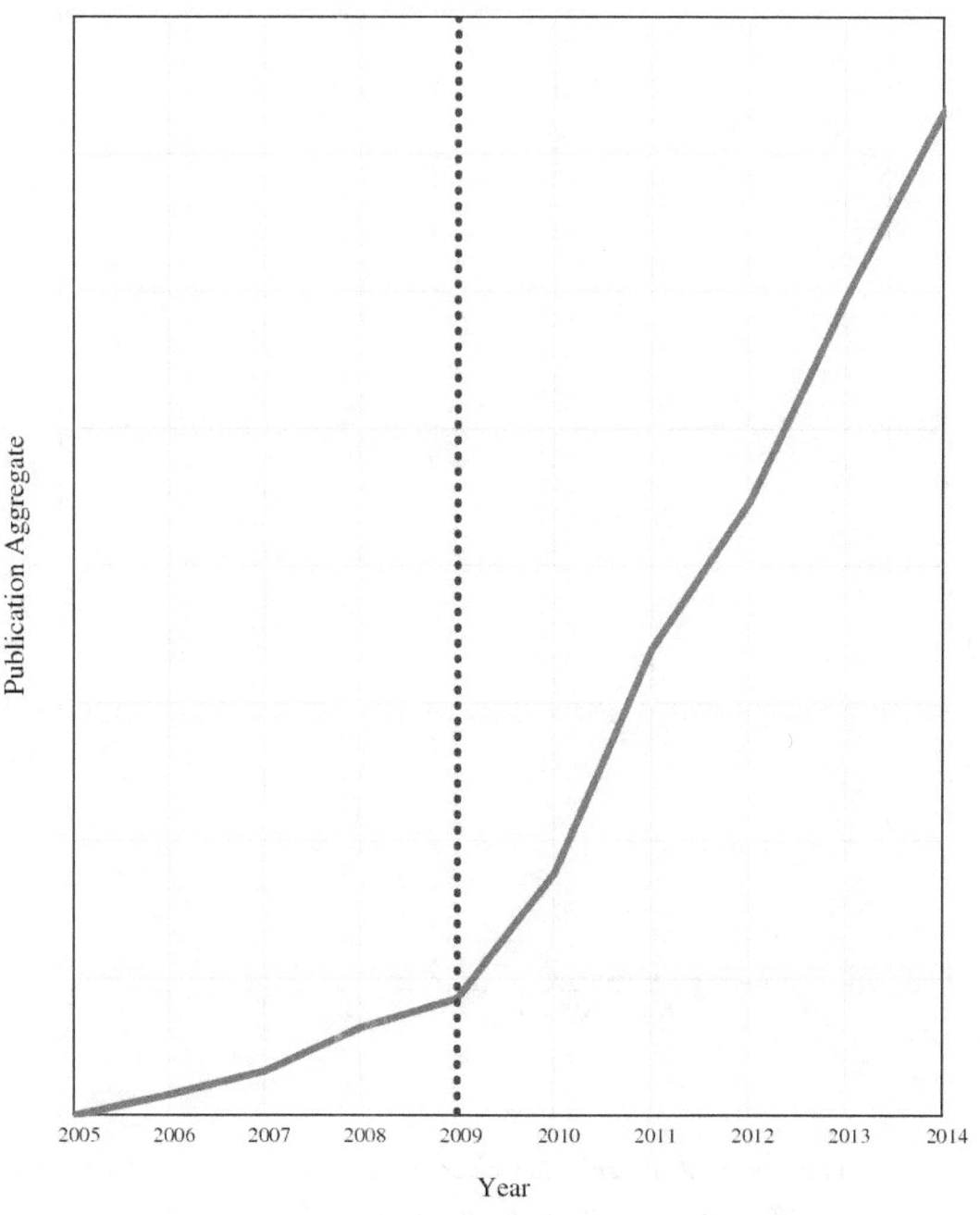

Figure 32 - Climate Change publication energizing event.

The difference here is that the event was not scientific. It was political. The date of the energizing event noted by the vertical dashed line is the beginning of Barack Obama's Presidency.

I was somewhat surprised to see this. I had assumed the motivation would have come from Vice President Al Gore's 2006 book and movie, *"An Inconvenient Truth."* But reality, as it often does, makes more sense. A motivational book has limited power to move a field of study, but funding can move mountains [27]. The Obama Administration directed approximately *$100 billion* dollars to climate change over his tenure, in one way or another.

The problem with politically energizing a field of scientific study, is that it has, by definition, a preconceived notion of what the truth is...even if it isn't. The funding will be applied in a manner consistent with the views of the funders and science is no longer the goal. The goal becomes to validate the preconceived notion and resolve whatever preconceived problem accompanies it.

The act of politically energizing a scientific field of study puts the political party and its members at risk if the field of study does not proceed in a manner consistent with the initial aims. As such, there is an intense, vested interest on behalf of the party and its members to ensure that the research yields results that, at least, do not put the party in a bad light.

To accomplish this, the party has great control over what research programs and researchers are funded. Further, once a political party attaches itself to a field of study, there will inevitably be institutions and individuals that have a partisan connection that motivates them to also guide events and results in order to strengthen the party. Not to mention the individuals who could care less about the party but have a corresponding preconceived view of what the science "should be."

The trend continues to the economic realm where business interests begin to change in order to maximize the business opportunities and status in the new environment. Businesses can, and will, adjust their practices, services, and products in order to enhance their bottom line. This is truly their job, and don't ever think they don't do it well. There need not be a partisan connection for this to be so. Just a change in the business environment.

For instance, if one looks at the Journal Nature, they have an incredibly well-respected reputation. They have an inspirational mission statement:

> First, to serve scientists through prompt publication of significant advances in any branch of science, and to provide a forum for the reporting and discussion of news and issues concerning science. Second, to ensure that the results of science are rapidly disseminated to the public throughout the world, in a fashion that conveys their significance for knowledge, culture and daily life. [28]

and a long, illustrious history beginning in 1869! But they are also a very well-run business.

In 2007, seeing the need, *Nature* started publishing *Nature Reports Climate Change* which was transitioned in 2011 to a full journal, *Nature Climate Change.* This was the first *Nature* publication to provide "...*research from social scientists*" alongside peer-reviewed "...*original research on climate change*" [29].

The transition of *Nature Reports Climate Change* to full-fledged journal status occurred just two years into the Obama Presidency. Considering the rate at which papers were being produced (as shown in **Figure 32**), it would have been negligent not to do so. But that increase in submitted papers was only possible because of the political decision to increase funding in the field.

The enormous power wielded by the Obama Administration when it came to funding climate research originated from their role in the granting process [30]. That is, the White House appoints the heads of the federal granting agencies. This had the effect to steer research and publication topics to those more sanitary to the political aims – even if the policy was unspoken, the agency head selection would be based on a common preconceived notion of the science involved.

So, the papers submitted to *Nature Climate Change* were rarely in the "denial" category (you don't get grants to prove the assumed science is wrong). And, based on the overwhelming volume of contrary results, *Nature* would be reticent to accept "denial" papers perceived as "already disproven".

Case in point, would be this book. Well, more accurately, the paper I submitted to *Nature* and many other major, scholarly, climate science related journals. This book is an extension of that paper. In 2015 I became interested in the whole "global warming" argument and quickly found much of what I've included herein. In 2016, I submitted *"Geothermal Contribution to Earth's Surface Temperature"* [31] to *Nature* and was surprised that it was rejected for "editorial" reasons. That is, it was never sent on for peer review.

I repeated this with other journals with exactly the same result. The paper was never passed on for peer review. It was always rejected for "editorial" reasons.

At the time, I was occupied with the development of a device to simulate an artificial-heart implanted patient, so I abandoned the paper assuming that it was just a matter of time before someone else made the geothermal discovery I had. Four years later, I was surprised to see that the error still had not been discovered. So, here we are.

I do not subscribe to any sort of conspiracy theory that scholarly journals were threatened with some sort of retaliation if "that Tinker guy's paper was published." Again, the vast bulk of work in the field would have been done under a granting agency motivated to reject proposals assumed to be contrary to preconceived reality. Journal editors would have seen these almost exclusively and anything contrary would have been viewed as "unlikely" if they were in generous mood.

That said, I have no patience and striking my head firmly against a wall is not my style. So, I'll not attempt publishing the findings of this book in a scientific journal again. Peer review will have to take place after the fact and in whatever public forum lends itself to the process.

So, the unintended consequence of granting agencies preferentially funding research in a manner prejudicial to work that does not support a preconceived notion can lead to censorship of that work far downstream of the granting agency itself. When you couple that with the vested interests of political and scientific partisans capable of publicly shaming those not of the same beliefs, the censorship becomes endemic. It is no wonder that my discovery was not repeated...or at least not published...in the intervening years.

So, now that we're here, what comes next? I'll take a stab at an answer in the next chapters.

Chapter 6 - Executive Summary

This chapter lists the numerous and significant challenges that must be addressed because of the flawed science proven in previous chapters. There are estimates that most scientists are of the opinion that the science is accurate and virtually all curricula reinforce the opinion that it is. Large expenditures have been, are being, and are planned to be made on behalf of taxpayers in order to protect against a threat that does not exist. Many suggested techniques have been proposed or are in place to mitigate the production of Greenhouse Gasses in the same effort. Some of these involve direct manipulation of Earth's upper atmosphere or other techniques to modulate insolation. Some of those techniques could pose significant risks. The chapter suggests that all of these points are detriments to society and will require a huge undertaking to eliminate.

It is important to understand that I have not proposed any new physics or mathematical techniques in the derivations of the preceding chapters. It is true that the application of the findings to Mercury orbit precession details was unexpected. But, at the core, all of what I have done should be rote to advanced students of physics. I would hope that the mistakes I uncovered would have denied students advancement and/or graduation. However, since these mistakes have been made almost exclusively by degreed or degree-track professionals, some with a multitude of academic awards, my hopes are moot.

It should be noted that, with the results of the preceding chapters, I have produced the most successful climate model in existence. Without a Greenhouse Effect, it accurately reproduces Earth surface temperatures present and historical. The historical temperatures so calculated reproduce all ice core and foram "temperature proxy" data as well as the abundances of trace gases formerly known as "Greenhouse Gasses". In this, the model is superior to any other climate model. Consequently, it nullifies the current definitions of climate change, climatology, global warming, and related concepts by proving that the common requirement of these topics, the Greenhouse Effect, does not exist.

As I see it, it is not an exercise in hyperbole to say that the beginning three chapters reveal possibly the largest and most costly collection of mistakes ever made in science. To begin with, researchers failed to successfully solve a routine partial differential equation, the heat equation in steady state, as shown in CHAPTER 1. This resulted in a violation of the first law of thermodynamics and the invention of an imagined process called the "Greenhouse Effect" to compensate for that mathematical mistake.

In testing the most viable theory, Milankovitch's [32] orbital forcing, the equation used to compute insolation was not only off by nearly two orders of magnitude, but it had the wrong sign. The researchers were actually viewing their graph upside down.

Finally, scientists calibrated isotope ratio abundance changes as a linear function of temperature when the time rate of change in temperature is dominant. This resulted in identifying events hundreds of thousands of years in error and labeling features as critical temperature spikes and valleys that were nothing more than variations in equilibrium state.

It only takes the first of these to shatter an entire field of study. Since the surface temperature of the planet is fully described by the combination of solar and geothermal flux, then there can be no such thing as a Greenhouse

Effect. Even if one wanted to be generous and allow for the possibility, the effect is too small to be measured.

In fact, CHAPTER 1 shows that the Greenhouse Effect itself was invented in order to explain a deficit in surface temperature that resulted from researchers failing to include geothermal flux in their calculations.

Sans the existence of a Greenhouse Effect, there can be no such thing as a Greenhouse Gas. Literally thousands of scientific papers have been written by climatologists detailing the intricacies of CO_2 and its deleterious effect on global temperatures. For these papers to be accurate, in any measure, relies on the existence of a Greenhouse Effect. Without that, the enormous expenditure in time, effort, and careers is all a complete waste.

Under the current definition, one that relies on the existence of a Greenhouse Effect, climatology as a field of study and endeavor does not exist. How many talented scientists, mathematicians, and programmers will have paid the ultimate career price for these mistakes before the debacle is over?

Astronomy does not go unscathed in this tragedy. Nasa, perhaps the most respected scientific entity on the planet, has invested incalculable man hours, computing resources, and space-based hardware to the study of Greenhouse Gas effects on today's environment [33] [34] -- an effect that does not exist. They, of all entities, should never have failed in the orbit calculations, but they did.

Carbon dioxide abundance is now used ubiquitously in geologic fields of study as a temperature forcer when it has no such role. It is not even a temperature proxy as shown in CHAPTER 3. The same is true of isotope abundance ratios in ice cores and forams. None of this data can be applied to discerning temperature.

No matter how bad it may seem, it is truly worse than can be imagined. Virtually every scientific and engineering organization [35] on the planet has made some sort of policy statement in solidarity to a climate change agenda that *requires* the existence of a Greenhouse Effect. These organizations are populated with members that are, presumably, the most well-trained individuals the world has to offer in finding flaws in science. They have not only failed in that, but they've compounded the problem by reinforcing the defective science.

Let us not forget those who have had to foot the bill for this disaster. Taxpayers the world over are poorly equipped to critically examine the esoteric processes that we do as scientists. They trust that we are professional enough, and have ample safety nets in place, that waste of their enormous effort and good will is jealously guarded against. They suffer our existence because, on average, we do more good than harm. They may have to rethink that.

And what of the politicians, heads of state, and the like? They have been advised, by respected scientists, that there exists a threat to the very planet on which we live. In good faith, they have responded by altering the fabric of our lives in order to combat a physical effect that does not exist. How can they trust us in the future? How can they unravel what has been done in the name of flawed science?

In attempting to thwart the predicted ravages of an erroneously theorized Greenhouse Effect, governments have rewarded certain industries with subsidies and preferential treatment [36] while punishing others with additional regulation and public scorn [37]. In this, they have created billionaires [36] on the largess of the taxpayer and weakened efficient industries that would prevail if allowed to compete on a level playing field. The cost of it all is borne by consumers and taxpayers who are understandably lacking in the training necessary to detect the scientific flaws or the ability to argue against the burden.

The courts are now involved in cases [38] where they are asked to assess responsibility for changes in environmental conditions and the resulting costs to address. The scientific community is the source of the arguments upon which these cases are based and will be the ultimate deciding factor in how they are resolved. Even the existence of these cases is an assault on justice systems if the underlying science is in error. To decide these cases in a way that depends on the existence of the Greenhouse Effect would be a fraudulent miscarriage of justice.

Despite the enormity already offered, there is one additional victim class that exceeds all others in scope. Virtually every human alive has been taught that the Greenhouse Effect and the resulting threat of Climate Change are real. As a result, the ability to understand this part of our universe has been stolen from everyone. We have put and are keeping the children of this world in a state of institutional ignorance all because of the failure of scientists to apply the physical laws and mathematical methods that they have been taught.

The truth is, I cannot see a resolution to the problem. What scientists have created in making these errors, is the largest and most invasive cult ever conceived. It is a religion that comes complete with the "priests" that made the mistakes in the first place and zealots who espouse the shunning, jailing, and any other manner of punishment of the unfaithful.

The tentacles of this church reach into every government, school, scientific institution, scientific society, periodical, news organization, political party, publisher, business, and playground. It has demanded and received at least a trillion dollars in offerings to date and is in the process of justifying an additional two to three trillion *per year* in order to combat devastation from a threat that does not exist.

There are serious recommendations and actual scientific studies where technology will be applied to modify the atmosphere or intercept solar flux [39] in order to slow or fix the problem. These recommendations from individuals that, by the findings herein, have already proven they do not understand the science or the mathematics involved.

There are sober attempts [40] to require businesses, governments, and citizens to pay dearly for the ability to manufacture goods, transport them and people, raise cows, drive cars, mow lawns, and on and on. All in the name of reducing the production of "Greenhouse Gasses" that do not exist. Requirements built on mistakes that should never have reached the public eye.

No matter how you look at it, there is no good news from this disaster. And it will take an enormous, concerted effort for the world to recover.

Chapter 7 - Executive Summary

In this chapter I will provide a few thoughts and suggestions for dealing with the issues that the discoveries in this book highlight. But the only true remedy is for the general public to decide if they will continue to support programs induced by flawed science.

For Everyone

- Evaluate what I've uncovered here. Convince yourself of its accuracy or not.
- If you don't understand the material in this book, I will be producing videos that will, hopefully, be less daunting. I'll make them accessible on the airofdoubt.com website as they become available.
- Study what I've reported here. Talk to professionals (scientists, mathematicians, engineers) and see if they can help you understand. Your local University should be invaluable. Ask them for advice on whether you should even believe what I've written. After all, if what I've written is correct, then scientists can make some incredible mistakes. And I'm a scientist.

For Scientists

- We will always make mistakes. In a perfect world, we catch the mistakes before they or their effects reach the public.
- In my opinion, the reason these mistakes were not intercepted before they became a problem was because political concerns biased funding in a way that kept science from doing its job.

- You should take steps in your own career to try to detect and refuse funding that has an obvious bias toward a particular conclusion. Even if, or especially if, that conclusion is one you share.

- The general public does not have the training that you do to evaluate scientific arguments that we make or even the language in which we make them. Critically evaluate what I've discovered here and, whatever opinion you have, see that you advise the people around you based on your best judgement.

- Whether you agree with this work or not, comment on it. Let your peers know. That is what we are supposed to do. Truth will out, but only if we are free to find it.

For Related Scientific Societies

- In my opinion, your reputation will, necessarily and justifiably, be diminished as a result of this. I would not fight that. Your value to the professionals you represent, and to the public that depends on those professionals, is greater than ever. Own up to the mistakes and get back to work.

- Your greatest contribution has always been to keep your membership motivated, engaged, and informed. All of that will be in great demand now. You have the tools and experience to guide your members to a bright and valuable future. Make that happen.

For the United Nations

- Arguably, the most visible champion of the Greenhouse Effect theory is the United Nations Intergovernmental Panel on Climate Change. This entity was created in 1988 with the intent to provide reliable, science-based information related to the perceived threat of inadvertent anthropogenic climate change. The organization does not carry out original research or climate monitoring but, instead, provides periodic reports gleaned from both scientific and anecdotal sources with commentary on their potential impact on climate and society. The reports support the United Nations Framework Convention on Climate Change (UNFCCC), the international treaty among member nations designed to "stabilize greenhouse gas concentrations in the atmosphere at a level that would prevent dangerous anthropogenic (human-induced) interference with the climate system."

- The findings of CHAPTER 1 serve to eliminate the need for the UNFCCC. The Greenhouse Effect is shown to be the result of the failure of scientists to properly account for the geothermal contribution to Earth's surface temperature. When properly incorporated in the relevant calculations, Earth's surface temperature is fully explained leaving the Greenhouse Effect at best infinitesimal or, more likely, simply invalid.

- The UN should make good on their promise to provide "reliable, science-based information" and, whether they are convinced that the information herein is valid or not, they should continue in that role appropriately.

For Scientific Journals

- It is proposed that the first three chapters of this book indicate that peer review journals have been ineffective in limiting the tenure of poor science.

- My own experience of not having papers subjected to peer review for editorial reasons indicates that non-scientists can and are interfering with the process of scientific advancement.

- It does not matter if the reasons for denying peer review are based on business concerns, prejudice, funding-forced prior art, ignorance, or mistake. The act itself is, by definition, censorship. And, when it results in damage to careers, education, government policy, and the very reputation of science (as in this case), the journals themselves must be reevaluated.
- Every peer-reviewed scientific journal should modify their policies and procedures in order to eliminate the possibility of bias, either intentional or incidental, in deciding which papers are submitted for peer review.
- If they are convinced that the discoveries documented here are valid, journals should review all papers they have published with an eye to determining if a paper is compromised by the material included herein. Those that have, should be prominently noted for future researchers. They should not be expunged since science gains as much on failed studies as it does on those that succeed.

For Governments

- As I showed in CHAPTER 5, you are capable of energizing fields of study through the power of funding. Review your procedures and make sure they guard against predetermining outcomes.
- The results I've given here indicate that your citizens are now ignorant of how a significant part of their physical universe works and they have been made so institutionally.
- Audit school programs to ensure they are not actively maintaining or extending the scope of misinformation.
- Ensure government programs are not misleading.
- Audit your laws, policies, publications, pamphlets, websites, videos, audible recordings, etc., to ensure invalid data is not being dispensed.

For Teachers

- This book details scientific mistakes that likely make some of what you are teaching, inaccurate.
- No matter which side of the fence you land on after evaluating what I've written, find a way to let the students know the truth. Another generation of ignorance would be a waste.

To the best of my ability, I have shown that serious and significant errors have been made in the general field of climate science by researchers of multiple disciplines. In doing so, I've created the most successful climate model in existence and the only classical mechanics-based model that successfully reproduces the Mercury perihelion precession rate.

The original data that showed a need for a Greenhouse Effect was a *33K* temperature deficit in calculations when compared with empirical data. I've shown that this exact value is found when applying the Stefan-Boltzmann equation to the *65mW/m²* geothermal flux. I have demonstrated that the standard mathematical technique known as the superposition principle demands that the geothermal flux be included in a manner that it contributes the full *33K* to Earth's surface temperature. As such, I have shown conclusively that the Greenhouse Effect does not exist and that it was an invention necessary to compensate for the mistake of not including geothermal flux in Earth surface temperature calculations in the first place. As there is no such thing as the Greenhouse Effect, there is also no such thing as a Greenhouse Gas.

I have further shown that researchers have derived and used a shockingly erroneous relationship to convert Earth orbit eccentricity into insolation estimates. I've provided a corrected formula that produces reliable and demonstrably accurate insolation and surface temperature estimates.

I have demonstrated that the isotopic abundance ratios found in forams and ice cores are not temperature proxies, as they have been used throughout climate science and related fields. They are, instead, proxies for the time rate of change of temperature. The corrected relationship between temperature and orbit eccentricity has been shown to accurately reproduce the ice core and foram proxy data; thereby validating the model and, simultaneously, proving that the Milankovitch Orbital Forcing Theory fully reproduces Earth's temperature history for at least the last million years.

Beyond reproducing the isotopic abundance ratio signals from forams and ice cores, I've reproduced CO_2 abundance data in the same manner. This demonstrates, both theoretically and graphically, that trace gas abundances in ice cores act merely as temperature-change monitors in the same manner as isotope abundance ratios do.

The solar system model that I have developed indicates that Kepler's Third Law may be made more accurate and useful. A factor including orbit eccentricity must be added to that law in order to correctly compare orbits with significantly different eccentricities. I have shown that the use of that additional factor resolves the Mercury Perihelion precession rate anomaly; a feat that no other classical mechanics-based analysis has done.

Heretofore, only the General Theory of Relativity was able to resolve the Mercury precession anomaly. Because of the success of the model derived herein accurately resolving the Mercury precession anomaly, I've shown that General Relativity is not needed to find that resolution. As such, General Relativity cannot claim a unique ability to do so.

The strength of these findings require the staggering conclusion that very little in the field of climate science is accurate. All climate models, of any level of sophistication or computing platform, use the isotopic abundance measurements as test data to prove fitness. As these have been proven to be temperature change proxies and not temperature proxies, no climate model aiming to reproduce them as temperatures have any hope of being accurate. Luckily, the model derived here will compensate.

As a result of all this, the challenges to governments, educators, scientific organizations, institutions of higher learning, and society itself are expected to be daunting in the immediate future. Simply the necessity to correct flawed teachings in a population of some six billion will be an enormous task. The cost of extracting society from this purgatory of ignorance will be borne by the same people who paid to get us here—the taxpayer. How that will be accepted is difficult to gauge.

I hope that the discoveries made and described here will eventually find use in the scientific community. Surely the truth can do that. But the near future promises to be one of abrupt and difficult change.

BIBLIOGRAPHY

[1] Q. Ma, "Greenhouse Gases: Refining the Role of Carbon Dioxide," [Online]. Available: https://www.giss.nasa.gov/research/briefs/ma_01. [Accessed 16 Nov 2019].

[2] D. E. Gray, Ed., American Institute of Physics Handbook, Third ed., New York: McGraw-Hill Book Company, 1972.

[3] A. Lacis, "Explaining climate," in *Our Warming Planet: Topics in Climate Dynamics*, C. Rosenzweig, D. Rind, A. Lacis and D. Manley, Eds., World Scientific - #EP 27, 2018, p. 3.

[4] P. Jones, M. New, D. Parker, S. Martin and I. Rigor, "Surface air temperature and its changes over the past 150 years," *Reviews of Geophysics,* vol. 37, no. 2, p. 173–199, 1999.

[5] H. N. Pollack, S. J. Hurter and J. R. Johnson, "Heat flow from the Earth's interior: Analysis of the global data set," *Reviews of Geophysics,* vol. 31, no. 3, pp. 267-280, 1993.

[6] Nasa, "ASTER Surface Emissivity AST05 Version 2.9," NASA - JPL, [Online]. Available: https://asterweb.jpl.nasa.gov/content/03_data/01_Data_Products/release_surface_emissivity_product.htm. [Accessed 25 09 2020].

[7] R. C. D. William E Boyce, Elementary Differential Equations and Boundary Value Problems, New York: John Wiley & Sons, 1977.

[8]   "Schlumberger Oilfield Glossary," Schlumberger, [Online]. Available: https://www.glossary.oilfield.slb.com/en/Terms/g/geothermal_gradi ent.aspx. [Accessed 13 10 2020].

[9]   H. Goldstein, Classical Mechanics, Reading, Massachusetts: Addison-Wesley Publishing Company, 1980.

[10] D. R. Williams, "Earth Fact Sheet," 22 April 2019. [Online]. Available: https://nssdc.gsfc.nasa.gov/planetary/factsheet/earthfact.html. [Accessed 29 November 2019].

[11] L. Graeme, L. Stephens, D. O'Brien, P. J. Webster, P. Pilewski and S. K. Jui-lin, "The albedo of Earth," *Reviews of Geophysics,* vol. 53, no. 1, pp. 141-163, 2015.

[12] BerkeleyEarth.org, R. Muller and R. Rhode, "Global Surface Temperatures: Best: Berkeley Earth Surface Temperatures," [Online]. Available: https://climatedataguide.ucar.edu/climate-data/global-surface-temperatures-best-berkeley-earth-surface-temperatures. [Accessed 03 10 2020].

[13] I. C. A. Organization, Manual of the ICAO Standard Atmosphere (extended to 80 kilometres (262 500 feet)) (Third ed.), 1993.

[14] G. A. Schmidt, "GISS Surface Temperature Analysis (GISTEMP) - FAQ," 14 01 2020. [Online]. Available: https://data.giss.nasa.gov/gistemp/faq/. [Accessed 03 10 2020].

[15] "Explaining Rapid Climate Change: Tales from the Ice," Nasa, 09 05 2006. [Online]. Available: https://earthobservatory.nasa.gov/features/Paleoclimatology_Evidence/paleoclimatology_evidence_2.php. [Accessed 13 10 2020].

[16] J. Lasker, E. Joutel and F. Boudin, "Orbital, precessional, and insolation quantities for the Earth from -20Myr to +10Myr," *Astronomy and Astrophysics,* vol. 270, pp. 522-533, 1993.

[17] S. B. N. Sharaf, *Trudy ITA,* vol. 11, no. 4, p. 231, 1967.

[18] W. Ward, "Climatic variations on Mars: 1. Astronomical theory of insolation," *Journal of Geophysical Research,* vol. 79, no. 24, pp. 3375-3386, 1974.

[19] A. Berger, "Long-Term Variations of Daily Insolation and Quaternary Climatic Changes," *Journal of Atmospheric Sciences,* vol. 35, no. 12, p. 2362–2367, 1978.

[20] J. M. Barnola, D. Raynaud, C. Lorius and N. I. Barkov, "Historical CO2 Record from the Vostok Ice Core (417,160 - 2,342 years BP)," January 2003. [Online]. Available: https://www.osti.gov/dataexplorer/biblio/dataset/1394139. [Accessed 26 09 2020].

[21] J. Petit, J. Jouzel, D. Raynaud and N. Barkov, "Climate and Atmospheric History of the Past 420,000 years from the Vostok Ice Core, Antarctica," *Nature, 399,* pp. 429-436, 1999.

[22] P. G. Mortyn, C. D. Charles, U. S. Ninnemann, K. Ludwig and D. A. Hodell, "Deep sea sedimentary analogs for the Vostok ice core," *Geochemistry Geophysics Geosystems,* vol. 4, no. 8, 2003.

[23] J. E. Tierney, J. Zhu, J. King, S. B. Malevich, G. J. Hakim and C. J. Poulsen, "Glacial cooling and climate sensitivity revisited," *Nature,* vol. 584, pp. 569-573, 2020.

[24] N. JPL, "Solar System Dynamics - HORIZONS Web Interface," NASA, [Online]. Available: https://ssd.jpl.nasa.gov/horizons.cgi. [Accessed 11 10 2020].

[25] C. Pollack, "University of Toronto Math Department, James Colliander, Math 426, Papers," 31 03 2003. [Online]. Available: http://www.math.toronto.edu/~colliand/426_03/Papers03/C_Polloc k.pdf. [Accessed 11 10 2020].

[26] W. Krätschmer, L. D. Lamb, K. Fostiropoulos and D. R. Huffman, "Solid C60: a new form of carbon," *Nature,* vol. 347, pp. 354-358, 1990.

[27] C. Flavelle, "To Protect Climate Money, Obama Stashed It Where It's Hard to Find," 15 03 2017. [Online]. Available: https://www.bloomberg.com/news/articles/2017-03-15/cutting-climate-spending-made-harder-by-obama-s-budget-tactics. [Accessed 28 09 2020].

[28] M. Statement, "About Nature," [Online]. Available: https://www.nature.com/nature/about. [Accessed 28 09 2020].

[29] O. Heffernan, "New beginnings," *Nature Climat Change,* vol. 1, p. 46, 2010.

[30] "Who's Who in the Federal Grant Policy-Making Community," [Online]. Available: https://www.grants.gov/web/grants/learn-grants/grant-policies/whos-who-in-federal-grant-policy.html. [Accessed 28 09 2020].

[31] F. A. Tinker, *Geothermal Contribution to Earth's Surface Temperature.,* Tucson, AZ, 2016.

[32] A. Buis, "Milankovitch (Orbital) Cycles and Their Role in Earth's Climate," Nasa, 27 02 2020. [Online]. Available: https://climate.nasa.gov/news/2948/milankovitch-orbital-cycles-and-their-role-in-earths-climate. [Accessed 13 10 2020].

[33] "What's NASA got to do with climate change?," Nasa, [Online]. Available: https://climate.nasa.gov/faq/18/whats-nasa-got-to-do-with-climate-change/. [Accessed 13 10 2020].

[34] J. Charles F. Bolden, "Message From the Administrator (764 Kb PDF) - NASA," 2011. [Online]. Available: https://www.nasa.gov/pdf/428442main_Message_from_the_Administrator.pdf. [Accessed 13 10 2020].

[35] D. Meerow, M. Stults and S. Stults, "Professional Societies and Climate Change," 01 2017. [Online]. Available: https://kresge.org/sites/default/files/library/env1007-psreport-0117_revised_11917.pdf. [Accessed 13 10 2020].

[36] J. Hirsch, "Elon Musk's growing empire is fueled by $4.9 billion in government subsidies," 30 03 2015. [Online]. Available: https://www.latimes.com/business/la-fi-hy-musk-subsidies-20150531-story.html. [Accessed 13 10 2020].

[37] N. Mulder, "https://www.thenation.com/article/archive/climate-green-new-deal/," The Nation, 10 11 2019. [Online]. Available: https://www.thenation.com/article/archive/climate-green-new-deal/. [Accessed 13 10 2020].

[38] D. Hasemyer, "Fossil Fuels on Trial: Where the Major Climate Change Lawsuits Stand Today," Inside Climate News, 17 01 2020. [Online]. Available: https://insideclimatenews.org/news/04042018/climate-change-fossil-fuel-company-lawsuits-timeline-exxon-children-california-cities-attorney-general. [Accessed 13 10 2020].

[39] P. Irvine, K. Emanuel, J. He, L. W. Horowitz, G. Vecchi and D. Keith, "Halving warming with idealized solar geoengineering moderates key climate hazards," *Nature Climate Change,* vol. 9, no. 4, p. 295, 2019.

[40] "Economics of Climate Change," US Environmental Protection Agency, [Online]. Available: https://www.epa.gov/environmental-economics/economics-climate-change. [Accessed 13 10 2020].